Created by Xspurts.com

All rights reserved.

Copyright © 2005 onwards .

By reading this book, you agree to the below Terms and Conditions.

Xspurts.com retains all rights to these products.

No part of this book may be reproduced in any form, by photostat, microfilm, xerography, or any other means, or incorporated into any information retrieval system, electronic or mechanical, without the written permission of Xspurts.com; exceptions are made for brief excerpts used in published reviews.

This publication is designed to provide accurate and authoritative information with regard to the subject matter covered but is for entertainment purposes only. It is sold with the understanding that the publisher is not engaged in rendering legal, accounting, health, relationship or other professional / personal advice. If legal advice or other expert assistance is required, the services of a competent professional should be sought.

♥ A New Zealand Designed Product

Get A Free Book At: https://free.xspurts.com

Table of Contents:

Table of Contents:
Gaia Philosophy: Introduction
Understanding Gaia Philosophy
Origin of Gaia Philosophy
Understanding Life and Nature
Balance of Nature
The Web of Life
Mother Earth: Gaia in Ancient Mythology
Gaia in Greek Mythology
Gaia in Other Cultures
The Gaia Hypothesis
Breakdown of The Gaia Hypothesis
Criticism and Reception
Gaia and Modern Science
Ecological Balance and Science
Gaia and Climate Change
Earth as a Single Organism
The Earth and Human Body Analogy
Humanity's Role in Gaia
Gaia and Spirituality
Pagan Beliefs and Gaia
Eastern Philosophies and Gaia
Gaia and Religion
Christianity and Gaia
Eastern Religions and Gaia
Environmental Ethics in Gaia Philosophy
Gaia and Sustainable Living
The Role of Balance in Ethics

The Concept of Gaia in Literature
Depictions of Gaia Across Literary Genres
Gaia in Science Fiction
Planet's Self-Regulation Mechanisms
The Earth's Feedback System
The Role of Biodiversity
Environmentalism and Gaia Philosophy
The Green Movement and Gaia
Ecological Activism Inspired by Gaia
Implementing Gaia Philosophy in Everyday Life
Eco-friendly Choices and Behaviours
Gaia-inspired Community Living
Gaia and the Future
Future Predictions Based on Gaia Philosophy
Gaia and Futuristic Technologies
Conclusion: Gaia Philosophy and its Global Impact
Implications for Science and Society
Our Responsibility Towards Mother Earth
Have Questions / Comments?
Get Another Book Free

Gaia Philosophy: Introduction

The concept of Gaia philosophy posits that the Earth and its biological systems behave as a single, self-regulating entity. Named after the Greek goddess Gaia, who personified the Earth, this philosophy intertwines ecological science with ethical considerations, emphasizing the interconnectedness of all living organisms and their environments. The roots of this perspective can be traced back to ancient philosophies, but it gained prominence in the late 20th century, largely due to the work of British scientist James Lovelock.

Lovelock introduced the Gaia hypothesis in the 1970s, suggesting that life on Earth actively contributes to the maintenance of conditions that sustain life. He argued that the biosphere, atmosphere, oceans, and soil are part of a complex system that regulates itself through feedback mechanisms. For instance, when carbon dioxide levels rise, plants may grow more vigorously, consuming more of the gas and helping to restore balance. This idea challenges the traditional view of nature as a collection of independent parts and highlights the importance of understanding the planet as an integrated whole.

The implications of Gaia philosophy extend beyond environmental science. It encourages a shift in perspective regarding humanity's role in nature. Rather than seeing humans as separate from or superior to other forms of life, this philosophy fosters a sense of stewardship. It prompts individuals and societies to consider how their actions impact the planet's health and emphasizes the responsibility to preserve the ecological balance.

In addition to its scientific foundations, Gaia philosophy has resonated with various spiritual and ethical movements. Many indigenous cultures have long recognized the interconnectedness of life, echoing the sentiments of Gaia. This alignment has led to a growing awareness of sustainability and environmental justice, inspiring movements advocating for the protection of ecosystems and the rights of all living beings.

Modern discussions around climate change, biodiversity loss, and environmental degradation often reference the principles of Gaia philosophy. The urgent need for collective action to combat these challenges reflects the interconnectedness emphasized by this perspective. By recognizing that human activities are part of a larger ecological system, individuals and communities are more likely to adopt practices that promote sustainability and resilience.

In summary, this holistic view of life on Earth underscores the importance of balance and cooperation within the ecosystem. As awareness of environmental issues grows, so does the relevance of Gaia philosophy, inspiring new generations to think critically about their relationship with the planet and its diverse forms of life.

Understanding Gaia Philosophy

Gaia philosophy embodies the idea that Earth functions as a cohesive, self-regulating system where all living and non-living components interact harmoniously. This perspective, rooted in ancient traditions, gained significant traction in the 20th century, largely attributed to the work of British scientist James Lovelock. He proposed the Gaia hypothesis, which posits that life on Earth contributes to maintaining conditions favorable for life, essentially viewing the planet as a single, complex organism.

Central to this philosophy is the concept of interconnectedness. Each element within the Earth's system, from microorganisms in the soil to vast forests and oceans, plays a vital role in sustaining the balance necessary for life. For example, the regulation of atmospheric gases, such as oxygen and carbon dioxide, is influenced by biological processes. Plants absorb carbon dioxide during photosynthesis and release oxygen, creating a cycle that supports both plant and animal life. Such interdependence challenges traditional views of nature as a mere collection of resources to be exploited and encourages a holistic understanding of ecological relationships.

Gaia philosophy extends its implications beyond ecology, influencing ethical considerations and human behavior. It invites a reevaluation of humanity's place in the natural world, advocating for a shift from domination to stewardship. This perspective emphasizes that human actions have far-reaching consequences on the planet's health and encourages individuals and societies to act responsibly to preserve the delicate balance of ecosystems.

Moreover, the philosophy has inspired movements centered on sustainability and environmental justice. Many indigenous cultures resonate with Gaia's principles, recognizing the sacredness of the Earth and the importance of living in harmony with nature. This alignment has fueled advocacy for protecting ecosystems, promoting biodiversity, and addressing climate change, highlighting the necessity of collective action to combat environmental degradation.

The relevance of this philosophy is increasingly apparent in contemporary discussions on global challenges. Climate change, loss of biodiversity, and pollution underscore the urgency of adopting practices that honor the interconnectedness of life. As societies grapple with these issues, Gaia philosophy serves as a guiding framework, inspiring a commitment to sustainable practices and a deeper appreciation for the intricate web of life on Earth.

By understanding and embracing the principles of Gaia, individuals are encouraged to reflect on their relationship with the planet, fostering a sense of responsibility that transcends generations. This awareness paves the way for a future where humanity coexists with the Earth's ecosystems, ensuring the well-being of all life forms and the preservation of our shared home.

Origin of Gaia Philosophy

The origins of Gaia philosophy can be traced back to a blend of ancient beliefs and modern scientific inquiry, particularly the contributions of British scientist James Lovelock in the 20th century. The name "Gaia" itself derives from Greek mythology, where Gaia is the personification of the Earth. This concept of Earth as a nurturing entity reflects a long-standing reverence for nature found in various cultures, emphasizing the interconnectedness of life.

The formal introduction of Gaia philosophy began with Lovelock's Gaia hypothesis, which he proposed in the 1970s. Lovelock posited that the Earth functions as a self-regulating system, where living organisms interact with their inorganic surroundings to maintain conditions conducive to life. This was a revolutionary idea that challenged the prevailing scientific views of the time, which often regarded ecosystems as separate and independent entities.

Lovelock's work was influenced by earlier thinkers, including the naturalist Charles Darwin, whose theories on evolution and natural selection laid the groundwork for understanding biological interdependence. Additionally, the ideas of other scientists, such as Lynn Margulis, who contributed to the concept of symbiosis, further shaped Lovelock's vision. Margulis's work on the role of microorganisms in maintaining ecological balance resonated with the principles of Gaia, reinforcing the notion that life plays an active role in regulating environmental conditions.

As Lovelock developed his hypothesis, he highlighted feedback loops within ecosystems, where biological processes influence atmospheric and geological conditions. For instance, the regulation of gases like carbon dioxide and oxygen is intricately linked to the activities of living organisms. This interplay underscores the idea that the Earth is not just a backdrop for life but an active participant in the survival of its inhabitants.

The emergence of Gaia philosophy coincided with growing environmental awareness in the late 20th century. As concerns about pollution, climate change, and habitat destruction gained prominence, the holistic perspective of Gaia offered a compelling framework for understanding the complexities of ecological interactions. The philosophy encouraged a paradigm shift in how humanity perceives its relationship with the Earth, promoting a sense of stewardship rather than exploitation.

In addition to its scientific roots, Gaia philosophy has resonated with various spiritual and ecological movements. Many indigenous cultures have long recognized the interconnectedness of life, echoing the principles of Gaia. This alignment has fueled advocacy for sustainability and environmental justice, inspiring efforts to protect ecosystems and promote biodiversity.

Today, the origins of Gaia philosophy serve as a foundation for ongoing discussions about our role in the natural world. As contemporary challenges such as climate change and biodiversity loss intensify, the philosophy provides valuable insights into fostering a harmonious relationship with the Earth. By recognizing the intricate web of life that sustains us, Gaia philosophy encourages a deeper understanding of the responsibility we hold in preserving the planet for future generations.

Understanding Life and Nature

At the heart of Gaia philosophy lies the profound understanding of life and nature as interconnected and interdependent systems. This perspective emphasizes that all living organisms, from the smallest microorganisms to the largest mammals, exist within a complex web of relationships that sustain life on Earth. Rather than viewing nature as a collection of separate entities, this philosophy encourages an appreciation for the intricate connections that bind them together.

The concept asserts that life influences the environment and vice versa. For example, plants play a crucial role in regulating atmospheric gases, absorbing carbon dioxide during photosynthesis and releasing oxygen. This symbiotic relationship not only supports plant growth but also ensures a breathable atmosphere for animals and humans. Such feedback mechanisms highlight how living organisms actively participate in shaping their surroundings, reinforcing the idea that Earth functions as a self-regulating system.

Understanding this dynamic interdependence fosters a sense of responsibility towards the environment. As humans, our actions can significantly impact ecosystems, and recognizing this connection calls for a more sustainable approach to living. The philosophy advocates for practices that prioritize ecological balance, urging individuals and communities to consider the long-term effects of their decisions on the planet. This shift in mindset encourages stewardship, where the goal is to protect and nurture the Earth rather than exploit its resources.

The implications of this philosophy extend beyond environmental science into ethical considerations. By acknowledging that all life forms are part of a greater whole, Gaia philosophy promotes compassion and respect for the natural world. It encourages us to reflect on our role within this system, fostering a sense of empathy for other living beings. This mindset can lead to actions that support biodiversity and conservation efforts, ultimately contributing to the health of the planet.

Additionally, Gaia philosophy resonates with various cultural and spiritual traditions that recognize the sacredness of nature. Many indigenous cultures have long understood the importance of living in harmony with the Earth, emphasizing the need for balance and respect for all forms of life. This alignment with Gaia's principles has inspired movements advocating for environmental justice, highlighting the rights of nature and the need for sustainable practices that benefit both people and the planet.

As contemporary challenges like climate change and habitat destruction intensify, the insights provided by Gaia philosophy become increasingly relevant. The interconnectedness of life calls for collective action to address these issues, encouraging collaboration across disciplines and communities. By understanding life and nature through the lens of this philosophy, we can work towards creating a sustainable future that honors the delicate balance of our ecosystems and recognizes the intrinsic value of all living beings.

Balance of Nature

The balance of nature is a foundational principle within Gaia philosophy, highlighting the intricate relationships and interdependencies that exist among all living organisms and their environments. This concept asserts that ecosystems are dynamic systems maintained through complex feedback loops, where each element plays a vital role in sustaining the overall health of the planet. Understanding this balance is crucial for recognizing how disruptions can lead to cascading effects that impact both local and global ecosystems.

At its core, the balance of nature involves the interplay between various biotic (living) and abiotic (non-living) components. For instance, predator-prey relationships are essential in regulating population sizes. If a predator species thrives due to an abundance of prey, it can lead to overhunting, which may subsequently decrease the prey population. This decline can create a ripple effect, impacting other species that rely on the prey for survival, thus altering the entire ecosystem. Such interconnectedness emphasizes that every action, whether by humans or other organisms, can have far-reaching consequences.

Furthermore, Gaia philosophy posits that life itself contributes to the stability of these systems. Organisms modify their environment through biological processes, creating conditions that favor life. For example, plants not only provide oxygen and food but also contribute to soil formation and moisture retention. Wetlands act as natural buffers, filtering pollutants and providing habitats for various species. These contributions showcase how life actively participates in maintaining ecological balance.

Human activity poses significant challenges to this natural equilibrium. Deforestation, pollution, and climate change are just a few examples of how our actions can disrupt established systems, leading to biodiversity loss and altered habitats. The degradation of ecosystems not only threatens the species that inhabit them but also undermines the services that these systems provide to humanity, such as clean water, fertile soil, and climate regulation.

Recognizing the balance of nature within the framework of Gaia philosophy calls for a reevaluation of how we interact with our environment. It emphasizes the need for sustainable practices that promote ecological health. By fostering biodiversity, restoring habitats, and reducing pollution, we can support the natural balance that sustains life on Earth.

Moreover, this philosophy encourages a holistic approach to environmental stewardship, advocating for the integration of ecological principles into decision-making processes. This perspective aligns with movements for environmental justice, where the rights of nature are respected, and the impacts of human activities on ecosystems are carefully considered.

Ultimately, understanding the balance of nature through the lens of Gaia philosophy inspires a sense of responsibility and interconnectedness. It encourages individuals and communities to recognize their role within the broader ecological system, promoting actions that honor the intricate web of life. By doing so, we can work towards a sustainable future that respects the balance essential for the well-being of all living organisms on our planet.

The Web of Life

At the heart of Gaia philosophy lies the intricate concept of the web of life, which emphasizes the interconnectedness and interdependence of all living organisms and their environments. This perspective illustrates that every species, from the tiniest microorganisms to the largest mammals, plays a crucial role in sustaining the ecosystem. The web of life signifies that changes to one component can reverberate throughout the entire system, affecting countless others.

The idea of interconnectedness in nature is rooted in various ecological principles. For instance, food webs illustrate how energy flows through an ecosystem. Producers, such as plants, convert sunlight into energy through photosynthesis, forming the base of the food chain. Herbivores consume these plants, and in turn, carnivores prey on herbivores. Each organism relies on others for survival, showcasing the delicate balance maintained within this web. If one species were to decline or become extinct, the effects could cascade, leading to overpopulation of some species and the decline of others, ultimately destabilizing the entire ecosystem.

Microbial life also plays an essential role in this interconnected system. Bacteria and fungi decompose organic matter, recycling nutrients back into the soil and making them available for plants. This process not only sustains plant life but also supports the animals that depend on plants for food. The relationships between different organisms often extend beyond direct interactions; for example, plants can communicate with each other through chemical signals, alerting neighbors to threats such as pests. This level of interaction highlights the complexity and sophistication of ecological networks.

Understanding the web of life within the framework of Gaia philosophy fosters a sense of responsibility toward environmental stewardship. It emphasizes that human actions can significantly impact this delicate balance. Habitat destruction, pollution, and climate change are examples of how human activities disrupt these interconnected systems. As ecosystems are altered or destroyed, the consequences can lead to loss of biodiversity and degradation of essential ecological services, such as clean water, air, and fertile soil.

Moreover, this philosophy encourages us to adopt a holistic view when considering environmental policies and practices. By recognizing that all life is interconnected, we are prompted to consider the broader implications of our actions. Sustainable practices, such as reforestation, conservation of natural habitats, and reduction of waste, become critical for maintaining the integrity of the web of life.

In addition, the web of life resonates with various cultural beliefs that emphasize harmony with nature. Many indigenous cultures have long understood the significance of these relationships, fostering a deep respect for the natural world. This perspective aligns with the principles of Gaia, encouraging a more profound appreciation for the roles of all organisms within the ecosystem.

Ultimately, embracing the concept of the web of life within Gaia philosophy inspires a collective commitment to protect and preserve the intricate relationships that sustain our planet. By fostering a sense of interconnectedness, we can work towards a future where humanity coexists harmoniously with nature, ensuring the survival and flourishing of all living beings.

Mother Earth: Gaia in Ancient Mythology

In ancient mythology, the figure of Gaia represents the embodiment of the Earth, revered as a nurturing and life-giving force. This archetype is a central figure in Greek mythology, symbolizing fertility, growth, and the interconnectedness of all life forms. Gaia's role extends beyond mere personification; she is often depicted as the source of all creation, emphasizing the belief in a harmonious relationship between humanity and nature.

Gaia emerged at the dawn of creation, born from Chaos, the primordial void. In mythological narratives, she is the mother of several significant deities, including Uranus (the sky), who together produced the Titans, powerful gods that ruled before the Olympians. This lineage underscores her status as a foundational figure in the pantheon of Greek mythology, representing the interconnectedness of earth and sky and the natural order of the universe.

Her significance in mythology illustrates a profound respect for the Earth and its life-sustaining properties. Gaia was worshipped as a goddess of fertility, often associated with agriculture and the abundance of the harvest. Ancient civilizations recognized the importance of maintaining a balanced relationship with nature to ensure prosperity and survival. Rituals and offerings were made in her honor, reflecting an understanding of the delicate balance that sustains ecosystems.

Gaia's mythology also carries a cautionary aspect. In the myths, her relationship with Uranus turned tumultuous when he imprisoned their children, the Titans, within her. In response, Gaia conspired with her son Cronus to overthrow Uranus, highlighting themes of rebellion and the struggle for balance in nature. This narrative emphasizes the idea that disruption in relationships, whether among deities or within ecosystems, can lead to chaos and disorder.

The reverence for Gaia in ancient mythology mirrors contemporary interpretations found within Gaia philosophy. Both perspectives emphasize the interconnectedness of life and the importance of maintaining harmony with the natural world. Modern environmental movements often draw inspiration from these ancient beliefs, advocating for sustainable practices that honor the Earth as a living entity deserving of respect and care.

Additionally, the notion of Gaia as "Mother Earth" resonates across various cultures. Many indigenous peoples worldwide view the Earth as a maternal figure, nurturing and sustaining all life. This universal theme underscores a collective acknowledgment of the Earth's importance and the responsibilities that come with it.

In essence, the legacy of Gaia in ancient mythology serves as a powerful reminder of the enduring bond between humanity and the natural world. As society grapples with environmental challenges today, revisiting these ancient narratives can inspire a renewed commitment to fostering harmony with the Earth, echoing the wisdom of those who revered Gaia as a life-giving force. Through understanding and respecting this connection, humanity can strive toward a more sustainable future that honors the intricate web of life that sustains us all.

Gaia in Greek Mythology

In Greek mythology, Gaia is a primordial goddess representing the Earth, embodying both the land and the life it nurtures. As one of the first entities to emerge from Chaos, the formless void, Gaia is revered as the mother of all living things. This deep connection to the Earth establishes her as a foundational figure within the Greek pantheon and a symbol of fertility, growth, and the intricate balance of nature.

Gaia's lineage is significant; she is the mother of Uranus (the sky), from whom she bore the Titans, a race of powerful deities who ruled before the Olympian gods. Among her children were notable figures such as Cronus and Rhea, who played crucial roles in the mythology surrounding the rise of the Olympians. The narratives surrounding Gaia highlight her vital role in the creation and sustenance of life, underscoring the interconnectedness that is central to Gaia philosophy.

In various myths, Gaia is depicted as a nurturing mother who provides for her children. However, her relationship with Uranus reveals the darker side of creation. When Uranus imprisoned their offspring, the Titans, within her, Gaia was driven to rebellion. She conspired with her son Cronus to overthrow Uranus, leading to a significant shift in power within the mythological landscape. This act not only emphasizes Gaia's agency but also reflects the delicate balance between creation and destruction inherent in natural processes.

The worship of Gaia was integral to ancient Greek culture, with rituals and offerings made in her honor to ensure fertility and abundance. Her image as Mother Earth resonated deeply with agricultural societies that relied on the land for sustenance. People recognized the importance of maintaining a harmonious relationship with the Earth to foster prosperity, reflecting themes that resonate with contemporary environmental ethics found in Gaia philosophy.

In addition to her nurturing role, Gaia is also associated with the prophetic. In some myths, she was revered as a source of wisdom, offering guidance and insights to both gods and mortals. This connection to foresight further underscores her importance as a guiding force within the natural world, reminding humanity of the consequences of their actions.

The legacy of Gaia in Greek mythology continues to inspire modern interpretations and environmental movements. Her embodiment of the Earth as a living entity reflects a growing recognition of the need to respect and care for our planet. The principles of Gaia philosophy emphasize the interconnectedness of life, urging individuals to consider the ecological consequences of their actions and to adopt a stewardship role toward the Earth.

Ultimately, Gaia's mythological narrative serves as a powerful reminder of the intricate relationships that define our existence. By embracing the wisdom inherent in these ancient stories, we can cultivate a deeper appreciation for the Earth and the life it supports, fostering a more sustainable future that honors the spirit of Gaia.

Gaia in Other Cultures

The concept of a mother figure representing the Earth and embodying life is a common thread in various cultures worldwide, echoing the principles of Gaia philosophy. Many civilizations have revered the Earth as a nurturing entity, highlighting the interconnectedness of all living beings and their environments. These cultural interpretations often reflect a deep respect for nature and a recognition of humanity's place within the broader ecological system.

In ancient Egyptian mythology, the goddess Nut personified the sky and was often depicted arching over the Earth, represented by her counterpart Geb. Together, they symbolize the unity of the heavens and the Earth, emphasizing the interconnectedness of life. Nut's role as a mother figure highlights the importance of fertility and regeneration, akin to the reverence shown to Gaia in Greek mythology.

Similarly, in Hinduism, the Earth is personified as the goddess Bhumi or Prithvi, who is celebrated as a nurturing mother. She is often invoked in rituals and prayers for agricultural prosperity and environmental harmony. This reverence for the Earth is integral to Hindu philosophy, which emphasizes the need to live in balance with nature, recognizing it as a sacred entity deserving of respect and protection.

Indigenous cultures also reflect Gaia's principles through their spiritual beliefs and practices. Many Native American tribes, for instance, view the Earth as a living being, often referring to it as "Mother Earth." This perspective fosters a deep connection to the land, emphasizing stewardship and sustainability. The Lakota Sioux, for example, have a saying, "Mitakuye Oyasin," which translates to "all my relatives," reinforcing the idea that all living beings are interconnected and part of a larger family.

In the African tradition, the goddess Gaia is often paralleled by deities such as Nzambi or Mbombo, who are associated with creation and the Earth. These figures embody the nurturing aspects of nature and the vital role of the Earth in sustaining life. Rituals and ceremonies honoring these deities often involve gratitude for the land and a commitment to living harmoniously with the environment.

In more contemporary contexts, the Gaia philosophy has inspired various ecological movements globally. The awareness of climate change, biodiversity loss, and environmental degradation has prompted many to adopt the principles found in these ancient cultures, advocating for a sustainable relationship with the Earth. The recognition that human actions have profound impacts on the environment resonates strongly with the teachings of many traditional societies that emphasize harmony and balance.

The parallels between Gaia and these cultural representations underscore a universal understanding of the Earth's importance and the need for stewardship. By honoring these diverse perspectives, humanity can cultivate a greater appreciation for the interconnectedness of all life, fostering a sense of responsibility to protect and sustain the planet for future generations. This shared wisdom reinforces the idea that caring for the Earth is not merely an ecological concern but a profound ethical imperative that transcends cultural boundaries.

The Gaia Hypothesis

The Gaia Hypothesis, formulated by British scientist James Lovelock in the 1970s, posits that Earth functions as a self-regulating system, where living organisms and their inorganic surroundings interact to maintain conditions that support life. This revolutionary idea emerged during a time when environmental concerns were becoming increasingly prominent, providing a scientific framework to understand the complex relationships within ecosystems.

At the core of this hypothesis is the assertion that life on Earth plays a crucial role in regulating the planet's environment. Lovelock proposed that biological processes, such as photosynthesis and respiration, contribute to maintaining stable conditions, including temperature, atmospheric composition, and ocean salinity. For instance, as carbon dioxide levels rise, plants absorb more of this greenhouse gas, thereby regulating its concentration in the atmosphere. This feedback loop exemplifies the dynamic interplay between life forms and their environment, highlighting the interconnectedness central to Gaia philosophy.

One of the key aspects of the Gaia Hypothesis is its emphasis on feedback mechanisms. These processes ensure that changes within the ecosystem trigger responses that help restore balance. For example, when global temperatures rise, ice caps may melt, leading to changes in ocean circulation patterns. This, in turn, affects climate and weather systems, demonstrating how shifts in one area can have widespread implications. By framing Earth as a living entity that actively maintains its conditions, Lovelock challenges traditional views of nature as merely a collection of separate parts.

The Gaia Hypothesis also encourages a shift in perspective regarding humanity's role on the planet. Rather than viewing ourselves as separate from or dominant over nature, it suggests that we are integral to the Earth's systems. This understanding fosters a sense of responsibility to protect and preserve the delicate balance that sustains all life. The implications of this perspective have significant ethical and ecological ramifications, urging societies to adopt sustainable practices that honor the interconnectedness of life.

Over the years, the Gaia Hypothesis has sparked extensive scientific research and discussion, contributing to fields such as ecology, environmental science, and climate studies. It has provided a framework for understanding complex systems and their responses to external pressures, particularly in the context of climate change. While some aspects of the hypothesis have been met with skepticism, its core principles have garnered widespread acceptance and have influenced a range of environmental movements.

The enduring relevance of the Gaia Hypothesis lies in its ability to inspire a deeper appreciation for the planet and its ecosystems. It emphasizes the importance of recognizing the intricate relationships that exist among all living beings and their environments. As contemporary society grapples with pressing environmental challenges, the insights provided by the Gaia Hypothesis offer valuable guidance for fostering a sustainable future, underscoring the interconnected nature of life on Earth and the urgent need for collective action to protect it.

Breakdown of The Gaia Hypothesis

The Gaia Hypothesis encompasses several key principles that articulate how life on Earth interacts with its environment to maintain conditions conducive to life. At its foundation is the idea that living organisms, including plants, animals, and microorganisms, play an active role in regulating their environment. This interdependence forms a complex web of relationships essential for sustaining life.

One of the fundamental concepts within the hypothesis is the notion of homeostasis, which refers to the ability of a system to maintain stable internal conditions despite external changes. Lovelock proposed that the Earth's biosphere functions similarly to a living organism, where various biological processes work together to create equilibrium. For instance, the regulation of gases in the atmosphere, such as oxygen and carbon dioxide, is influenced by the activities of living organisms. Through photosynthesis, plants absorb carbon dioxide, while respiration in animals releases it, showcasing a dynamic feedback mechanism that helps stabilize atmospheric composition.

Another critical aspect of the Gaia Hypothesis is the idea of feedback loops. These loops illustrate how changes in one part of the ecosystem can lead to responses that restore balance. For example, if global temperatures rise due to increased greenhouse gas emissions, this can lead to the melting of polar ice caps. As ice melts, it decreases the Earth's albedo (reflectivity), causing more solar energy to be absorbed, which further raises temperatures. This feedback effect highlights the interconnectedness of various components within the Earth's system and emphasizes the delicate balance that must be maintained.

The Gaia Hypothesis also suggests that Earth's systems are self-organizing. Living organisms adapt and evolve in response to environmental changes, contributing to the resilience of the ecosystem. This adaptability is crucial for survival, allowing species to respond to shifts in climate, resource availability, and other pressures. For instance, the proliferation of certain plant species in response to changing rainfall patterns can influence local ecosystems, further illustrating the dynamic nature of life on Earth.

Moreover, the hypothesis emphasizes humanity's role within this intricate system. Rather than being separate or superior to nature, humans are viewed as integral participants in the Earth's biosphere. This perspective fosters a sense of stewardship and responsibility, urging individuals and societies to consider the environmental consequences of their actions. The recognition that human activities can disrupt natural balance reinforces the importance of sustainable practices that promote ecological health.

While the Gaia Hypothesis has faced skepticism, particularly concerning its implications about intentionality or purpose in nature, it has significantly influenced environmental science and ecological thought. It has prompted further research into complex systems, resilience theory, and climate science, providing a framework for understanding how life interacts with and shapes the planet.

In summary, the Gaia Hypothesis illustrates the intricate connections between life and the environment, highlighting the mechanisms that sustain ecological balance. By emphasizing the interdependence of all living organisms, it calls for a reevaluation of humanity's relationship with the Earth, advocating for a more sustainable and harmonious existence within the web of life.

Criticism and Reception

The Gaia Hypothesis, while influential and thought-provoking, has faced criticism and skepticism from various quarters, particularly within the scientific community. Critics argue that it can anthropomorphize the Earth, attributing intentions and purpose to natural processes that operate without conscious direction. This perspective raises questions about the validity of viewing the Earth as a living organism with its own goals, which some scientists find problematic. They contend that this viewpoint could lead to misunderstandings about ecological dynamics and the complexities of biological systems.

Another significant criticism revolves around the hypothesis's testability. Some scientists argue that the Gaia Hypothesis lacks empirical rigor, as it is difficult to design experiments that definitively prove or disprove the concept. While Lovelock and his supporters have presented various observations supporting the idea of self-regulation in ecosystems, detractors point out that correlation does not necessarily imply causation. They argue that many ecological processes can be explained through established scientific principles without invoking the idea of Gaia as a self-regulating system.

Despite these criticisms, the Gaia Hypothesis has found a receptive audience, particularly among environmentalists and those advocating for sustainability. The philosophy resonates with movements that emphasize the interconnectedness of life and the importance of preserving ecological balance. Its holistic approach encourages individuals and societies to consider the broader implications of their actions on the environment, fostering a sense of responsibility toward the planet.

The hypothesis has also inspired interdisciplinary research, bridging gaps between ecology, Earth sciences, and even social sciences. Scholars have explored the connections between ecosystems and human behavior, leading to a greater understanding of how societal choices impact the environment. This dialogue has spurred initiatives focused on climate change, biodiversity conservation, and sustainable practices, further embedding the principles of Gaia philosophy in contemporary environmental discourse.

Additionally, the Gaia Hypothesis has influenced cultural and artistic expressions, reflecting a growing awareness of the need to coexist harmoniously with nature. Literature, art, and music inspired by Gaia themes often emphasize the beauty of interconnectedness and the importance of protecting the planet.

Over the years, the discussion surrounding the Gaia Hypothesis has evolved. While initial critiques raised valid concerns about its scientific foundations, ongoing research and dialogue have enriched our understanding of ecological systems. The hypothesis serves as a springboard for exploring complex relationships within the biosphere and emphasizes the need for a paradigm shift in how humanity perceives its role in the natural world.

In conclusion, the Gaia Hypothesis has sparked significant debate and discussion, balancing criticism with a broader acceptance that highlights the urgency of environmental stewardship. As contemporary society confronts pressing ecological challenges, the ideas embedded in Gaia philosophy continue to inspire efforts toward sustainability and a deeper appreciation for the intricate web of life on Earth.

Gaia and Modern Science

The relationship between Gaia philosophy and modern science has sparked intriguing discussions about the interconnectedness of life and the planet's ecological systems. The Gaia Hypothesis, proposed by James Lovelock, posits that Earth operates as a self-regulating entity, where living organisms interact with their environments to maintain conditions conducive to life. This perspective aligns with various scientific fields, including ecology, climatology, and Earth systems science, and has opened new avenues for research and understanding.

One of the critical areas where Gaia philosophy intersects with modern science is in the study of feedback mechanisms within ecosystems. Scientists have increasingly recognized that ecological processes are highly interconnected and can exhibit self-regulating behaviors. For instance, the relationship between carbon dioxide levels and plant growth exemplifies a feedback loop. Plants absorb carbon dioxide during photosynthesis, and increased carbon levels can lead to enhanced plant growth, which in turn helps stabilize atmospheric conditions. This dynamic interplay reflects the core tenets of Gaia philosophy, emphasizing the active role of life in shaping the Earth's environment.

Moreover, advancements in Earth systems science have provided empirical support for aspects of the Gaia Hypothesis. Researchers use complex models to study how various components of the Earth system—atmosphere, oceans, land, and biosphere—interact and influence one another. These models demonstrate that changes in one system can lead to significant effects in others, reinforcing the notion of interconnectedness that Gaia philosophy espouses. For example, climate models have shown how deforestation and changes in land use can impact regional and global climates, illustrating the intricate relationships that define Earth's ecological balance.

The growing concern about climate change and environmental degradation has further highlighted the relevance of Gaia philosophy in contemporary scientific discourse. As scientists grapple with the implications of human activities on the planet, the idea that life can influence and regulate the environment has gained traction. This has led to a more integrated approach to environmental science, where researchers emphasize the need for sustainability and conservation efforts that respect the intricate relationships within ecosystems.

Additionally, the Gaia philosophy has inspired new interdisciplinary studies, combining insights from biology, ecology, geology, and atmospheric sciences. This collaborative approach has facilitated a more holistic understanding of ecological dynamics, encouraging scientists to explore how biological diversity contributes to ecosystem resilience. Such studies underline the importance of preserving biodiversity as a critical component of maintaining the Earth's self-regulating systems, resonating with the core principles of Gaia philosophy.

The philosophical implications of Gaia also extend to ethical considerations in science and environmental policy. By framing the Earth as a living system, Gaia philosophy encourages a shift in how humanity perceives its role within the biosphere. This perspective promotes a sense of stewardship, urging individuals and societies to adopt sustainable practices that prioritize ecological health. As a result, environmental ethics grounded in Gaia philosophy have influenced policies aimed at reducing carbon emissions, protecting habitats, and promoting sustainable resource management.

In conclusion, the dialogue between Gaia philosophy and modern science has enriched our understanding of the complex interrelationships that define life on Earth. As scientific research continues to explore the dynamics of ecological systems, the principles of Gaia offer valuable insights that emphasize the importance of interconnectedness and the need for responsible stewardship of the planet. This synergy between philosophy and science underscores a growing recognition of the delicate balance necessary for sustaining life in an increasingly complex and changing world.

Ecological Balance and Science

Ecological balance is a crucial concept within Gaia philosophy, emphasizing the intricate relationships and interdependencies among living organisms and their environments. This idea asserts that ecosystems are dynamic, self-regulating systems where various species interact with one another and their physical surroundings to maintain stability. Understanding this balance is essential for addressing contemporary environmental challenges and promoting sustainable practices.

At the core of ecological balance is the concept of biodiversity, which refers to the variety of life forms within a given habitat. High biodiversity contributes to ecosystem resilience, allowing systems to withstand disturbances and adapt to changes. For instance, a diverse range of plant species can enhance soil quality and prevent erosion, while varied animal populations can help regulate populations of pests and competitors. This interconnectedness aligns with the principles of Gaia philosophy, which emphasizes the role of life in maintaining the conditions necessary for survival.

Scientific research has increasingly highlighted the importance of ecological balance through various studies and ecological models. For example, the introduction of non-native species into an ecosystem can disrupt existing balances, leading to unforeseen consequences. In many cases, invasive species outcompete native flora and fauna, resulting in declines in biodiversity and the loss of ecosystem functions. Such studies demonstrate the delicate interplay between species and the environment, underscoring the need for conservation efforts aimed at preserving native species and habitats.

Another key aspect of ecological balance is the role of trophic levels in food webs. These levels illustrate how energy flows through ecosystems, with producers at the base (such as plants) supporting herbivores and, subsequently, carnivores. Disruptions at any level can have cascading effects throughout the food web. For instance, the decline of a top predator can lead to an overabundance of herbivores, which may then overgraze vegetation, affecting the entire ecosystem's structure. This interconnectedness highlights the need for a holistic approach to environmental management, as changes in one component can significantly impact others.

Climate change poses a significant threat to ecological balance, as it alters the conditions that ecosystems depend on for stability. Rising temperatures, shifting precipitation patterns, and increased frequency of extreme weather events can disrupt the delicate relationships among species and their habitats. For instance, coral reefs, which rely on specific temperature ranges, are increasingly threatened by ocean warming and acidification. The decline of coral ecosystems not only affects the myriad species that depend on them but also has broader implications for coastal protection and fisheries.

Addressing these challenges requires a multidisciplinary approach that incorporates insights from ecology, climatology, and social sciences. By understanding the principles of ecological balance within the framework of Gaia philosophy, scientists and policymakers can develop strategies that promote sustainability and resilience. This may include restoring degraded ecosystems, protecting biodiversity, and implementing policies that mitigate the impacts of climate change.

Moreover, fostering public awareness and appreciation of ecological balance is crucial for driving meaningful change. Education and community engagement can help individuals recognize their role within the ecosystem and the importance of preserving natural habitats. Initiatives that encourage sustainable practices, such as local conservation efforts and responsible resource use, can empower communities to actively participate in maintaining ecological balance.

In summary, the interplay between ecological balance and science is central to understanding the principles of Gaia philosophy. By recognizing the interconnectedness of life and the importance of biodiversity, we can address environmental challenges more effectively. Emphasizing sustainability and stewardship not only honors the complexity of ecosystems but also fosters a more harmonious relationship between humanity and the natural world.

Gaia and Climate Change

The interplay between Gaia philosophy and climate change highlights the intricate connections between living organisms and the Earth's systems. At its core, Gaia philosophy posits that the Earth functions as a self-regulating system where biological processes actively contribute to the maintenance of environmental stability. As climate change poses unprecedented challenges to this balance, understanding these dynamics becomes crucial for addressing the ecological crises facing our planet.

One of the central tenets of Gaia philosophy is the recognition that life influences the planet's climate and conditions. For instance, plants play a vital role in carbon dioxide absorption through photosynthesis, helping regulate atmospheric levels of greenhouse gases. However, human activities, such as deforestation and fossil fuel combustion, have significantly increased carbon emissions, disrupting the natural carbon cycle. This interference not only exacerbates climate change but also undermines the self-regulating mechanisms that Gaia philosophy advocates.

The effects of climate change are widespread and multifaceted, impacting ecosystems and biodiversity. Rising global temperatures, shifting precipitation patterns, and increasing frequency of extreme weather events threaten the delicate balance that sustains life. Coral reefs, for example, are highly sensitive to temperature changes; even slight increases can lead to bleaching and the collapse of these critical ecosystems. As species struggle to adapt to rapid changes, biodiversity loss becomes a pressing concern, further destabilizing ecosystems and diminishing their resilience.

Gaia philosophy emphasizes the interconnectedness of life, encouraging a holistic understanding of climate change. The cascading effects of rising temperatures can be seen in various ecosystems, where changes in one species can trigger shifts throughout the food web. For example, the decline of pollinators, such as bees, due to habitat loss and climate change can adversely affect plant reproduction, impacting entire food chains. This highlights the importance of maintaining biodiversity and ecological integrity in combating climate change.

Addressing climate change through the lens of Gaia philosophy calls for a shift in how humanity perceives its relationship with the Earth. Rather than viewing the planet as a resource to be exploited, this perspective promotes stewardship and responsibility. It encourages individuals and communities to adopt sustainable practices that honor the interconnectedness of all life forms. This may include reducing carbon footprints, advocating for renewable energy sources, and supporting conservation efforts that protect critical habitats.

The philosophy also inspires collective action and grassroots movements focused on environmental justice. As the impacts of climate change disproportionately affect marginalized communities, Gaia philosophy underscores the need for equitable solutions that prioritize the well-being of both people and the planet. Efforts to combat climate change must consider social, economic, and environmental factors, ensuring that all voices are heard in the pursuit of sustainability.

Moreover, Gaia philosophy promotes the idea of resilience—both in ecosystems and human societies. By fostering biodiversity and restoring degraded ecosystems, we can enhance the Earth's natural ability to adapt to changing conditions. This resilience is crucial for mitigating the effects of climate change and maintaining the ecological balance necessary for life.

In summary, the relationship between Gaia philosophy and climate change underscores the importance of recognizing the interconnectedness of life and the planet's systems. As climate change challenges the self-regulating mechanisms of Earth, embracing the principles of Gaia can guide us toward sustainable solutions that honor the delicate balance of our ecosystems. By fostering stewardship and collective action, we can work toward a future that supports both the health of the planet and the well-being of its inhabitants.

Earth as a Single Organism

The concept of Earth as a single organism is a foundational idea within Gaia philosophy, suggesting that all living and non-living components of the planet are interconnected and function as a cohesive, self-regulating system. This perspective views the Earth not merely as a collection of separate entities but as an integrated whole where biological processes contribute to the maintenance of environmental conditions conducive to life.

At the heart of this idea is the recognition of complex feedback mechanisms that operate within ecosystems. Just as the human body regulates its internal environment through various systems—such as temperature control, fluid balance, and waste elimination—Earth's biosphere performs similar functions. For example, the interplay between plants, animals, and microorganisms helps to regulate atmospheric gases, such as oxygen and carbon dioxide. Through photosynthesis, plants absorb carbon dioxide and release oxygen, contributing to a stable atmosphere essential for life.

Lovelock's Gaia Hypothesis posits that life actively influences the environment in ways that promote the conditions necessary for its own survival. This assertion emphasizes the importance of biodiversity and the roles that different species play within ecosystems. The presence of diverse organisms enhances resilience, allowing ecosystems to adapt to changes and recover from disturbances. For instance, coral reefs, which support an abundance of marine life, rely on the complex interactions between coral polyps, algae, and various fish species to maintain their health and functionality.

This holistic view extends beyond mere biology, incorporating geological and atmospheric processes into the understanding of Earth as a single organism. The cycling of nutrients, the regulation of climate, and the movement of water through the hydrological cycle are all integral components of this living system. Changes in one area—such as deforestation affecting rainfall patterns—can have cascading effects throughout the ecosystem, underscoring the interdependencies that characterize life on Earth.

The idea of Earth as a single organism also invites a reevaluation of humanity's role within this system. By acknowledging that humans are integral to the planet's health, this perspective encourages a sense of stewardship and responsibility. It challenges the notion of humans as separate from nature, advocating instead for a relationship characterized by respect and care. This shift in mindset is crucial in addressing contemporary environmental challenges, as unsustainable practices can disrupt the delicate balance that sustains life.

Moreover, this philosophy resonates with various cultural and spiritual traditions that view the Earth as a living entity. Many indigenous cultures have long recognized the interconnectedness of all life forms and the importance of living in harmony with the natural world. This alignment with Gaia philosophy reinforces the idea that a respectful relationship with the Earth is essential for both ecological and social well-being.

In the context of modern environmental science, the notion of Earth as a single organism has gained traction as researchers explore the implications of interconnected systems. Climate change, biodiversity loss, and habitat degradation illustrate how human activities can disrupt the delicate balance of the planet. Understanding these connections is critical for developing effective conservation strategies and promoting sustainability.

Ultimately, embracing the idea of Earth as a single organism fosters a deeper appreciation for the complexity and beauty of the natural world. By recognizing our place within this living system, we can work toward a more sustainable future that honors the interconnectedness of life and the Earth's ability to support it. This holistic perspective encourages collaborative efforts to protect and restore ecosystems, ensuring that the planet remains a vibrant and thriving home for all living beings.

The Earth and Human Body Analogy

The analogy between Earth and the human body provides a compelling framework for understanding Gaia philosophy and the interconnectedness of life. Just as the human body operates as a complex, self-regulating system comprised of various organs and tissues that work together to maintain health and functionality, Earth functions as a living organism, where its ecosystems and biogeochemical processes interact to sustain life.

In the human body, different systems—such as the circulatory, respiratory, and digestive systems—are interdependent. For example, the circulatory system delivers oxygen and nutrients to cells, while the respiratory system allows for the exchange of gases, providing oxygen to the bloodstream and removing carbon dioxide. This intricate collaboration ensures that all bodily functions are maintained and that the body can respond to changes in the environment, such as physical activity or illness.

Similarly, Earth's ecosystems serve as interconnected systems that support various forms of life. For instance, forests, wetlands, and oceans contribute to the planet's overall health by regulating climate, purifying water, and cycling nutrients. Forests act as lungs for the Earth, absorbing carbon dioxide and releasing oxygen through photosynthesis, much like the lungs in the human body. Wetlands filter pollutants and provide habitats for countless species, while oceans help regulate temperature and store carbon. Each of these ecosystems plays a vital role, and their health is essential for maintaining the balance of the planet.

The feedback mechanisms observed in both the human body and the Earth illustrate the principles of self-regulation central to Gaia philosophy. In the body, when a system is disrupted—such as when blood sugar levels rise after eating—the body activates mechanisms to restore balance, like insulin secretion to lower blood sugar levels. On a planetary scale, similar feedback loops exist; for example, increased carbon dioxide in the atmosphere can lead to enhanced plant growth, which, in turn, helps absorb more carbon dioxide, stabilizing the climate.

Human activities, however, can disrupt these natural processes, analogous to how disease can impact bodily functions. Pollution, deforestation, and climate change are examples of how human actions can impair the Earth's ability to self-regulate. Just as a disease can weaken the body and lead to a breakdown of its systems, environmental degradation can compromise ecosystems, resulting in biodiversity loss and diminished ecological resilience.

This analogy also underscores the importance of recognizing our role within this living system. Just as individuals must care for their bodies to maintain health, humanity has a responsibility to protect and nurture the planet. Gaia philosophy encourages a sense of stewardship, emphasizing that actions taken today can have profound implications for future generations and the overall health of the Earth.

By fostering an awareness of the interconnectedness of life, this analogy invites individuals to reflect on their choices and their impact on the environment. Sustainable practices—such as reducing waste, conserving energy, and protecting natural habitats—are akin to adopting a healthy lifestyle that promotes well-being in the body. When people understand their relationship with the Earth as part of a larger organism, they are more likely to engage in behaviors that support ecological balance and sustainability.

In conclusion, the Earth and human body analogy serves as a powerful illustration of Gaia philosophy, emphasizing the interconnectedness of life and the importance of maintaining balance within both systems. By recognizing that we are integral to the health of the planet, we can cultivate a deeper appreciation for the environment and a commitment to preserving the intricate web of life that sustains us all.

Humanity's Role in Gaia

Humanity plays a vital role within the framework of Gaia philosophy, which posits that Earth functions as a self-regulating, interconnected system where all living and non-living components work together to sustain life. This perspective challenges the notion of human exceptionalism, emphasizing instead that humans are integral parts of the Earth's ecological systems. Understanding this relationship fosters a sense of responsibility and stewardship toward the environment.

As agents of change, humans have a profound impact on the planet's ecosystems. Activities such as agriculture, urban development, and industrialization have transformed landscapes and altered natural processes. While these actions can lead to economic growth and technological advancement, they often come at the cost of environmental degradation, loss of biodiversity, and disruption of ecological balance. Recognizing the consequences of these actions is essential for fostering a more sustainable relationship with the Earth.

Gaia philosophy encourages individuals and societies to embrace their role as stewards of the environment. This involves adopting practices that promote ecological health and resilience. For instance, sustainable agriculture techniques, such as crop rotation, organic farming, and permaculture, aim to minimize environmental impact while ensuring food security. These methods not only help restore soil health but also support biodiversity, illustrating how responsible human actions can contribute positively to the planet's self-regulating systems.

In addition to sustainable practices, humanity's role extends to advocacy and education. Raising awareness about environmental issues is crucial for inspiring collective action and promoting a culture of sustainability. Grassroots movements and community initiatives can empower individuals to make informed choices and influence policies that prioritize ecological well-being. By fostering a sense of connection to the natural world, these efforts can cultivate a deeper understanding of the importance of preserving ecosystems for future generations.

The interconnectedness emphasized in Gaia philosophy also highlights the need for collaboration across disciplines and sectors. Climate change, habitat destruction, and pollution are complex issues that require holistic solutions. By integrating scientific research, community input, and policy-making, societies can develop strategies that address environmental challenges more effectively. Collaborative approaches, such as ecosystem-based management and participatory governance, can enhance resilience and promote sustainability at local, regional, and global scales.

Furthermore, humanity's role in Gaia philosophy underscores the ethical dimensions of environmental stewardship. The recognition that all life forms are interconnected calls for a shift in perspective regarding the intrinsic value of nature. This ethical framework encourages respect for the rights of non-human beings and the ecosystems they inhabit. Movements advocating for the protection of natural habitats and the rights of indigenous peoples exemplify this ethical commitment to preserving the planet's biodiversity.

In summary, the role of humanity within the context of Gaia philosophy is multifaceted, encompassing stewardship, advocacy, education, and ethical responsibility. By understanding that our actions significantly impact the planet's ecological balance, we can cultivate a more harmonious relationship with the Earth. Embracing this interconnectedness fosters a sense of collective responsibility, inspiring efforts to protect and restore ecosystems for the benefit of all living beings. Ultimately, acknowledging our role in the Gaia framework can lead to a more sustainable future, ensuring the health and vitality of the planet for generations to come.

Gaia and Spirituality

Gaia philosophy intertwines deeply with various spiritual beliefs, offering a holistic perspective on the interconnectedness of all life forms and the natural world. This philosophical framework presents Earth as a living entity, where every organism contributes to the overall health of the planet, mirroring many spiritual traditions that emphasize the sacredness of nature and the importance of maintaining harmony within it.

Many indigenous cultures have long recognized the Earth as a sacred being, often personifying it as a mother figure. This reverence for the planet is reflected in their spiritual practices, which typically involve rituals and ceremonies honoring the land, water, and all living creatures. Such beliefs resonate strongly with Gaia philosophy, which promotes the idea that the health of the Earth is intrinsically linked to the well-being of its inhabitants. By acknowledging the interconnectedness of all life, these spiritual traditions encourage stewardship and respect for the environment.

In addition to indigenous beliefs, contemporary spiritual movements have embraced Gaia philosophy, emphasizing the significance of ecological consciousness. Eco-spirituality, for example, integrates environmental concerns with spiritual practices, urging individuals to recognize their connection to the Earth and to engage in actions that promote ecological balance. This movement fosters a sense of responsibility towards nature, encouraging followers to see themselves as active participants in the Earth's ongoing processes.

The idea of Earth as a living organism also parallels various religious beliefs that highlight the sanctity of creation. In many faiths, the Earth is viewed as a reflection of a higher power, imbued with spiritual significance. For instance, in Hinduism, the Earth is revered as the goddess Bhumi, embodying fertility and abundance. Such beliefs encourage followers to care for the planet as a sacred duty, aligning closely with the principles of Gaia philosophy.

Furthermore, the recognition of the interconnectedness of life fosters a deeper understanding of humanity's role within the larger ecological system. This perspective encourages individuals to cultivate a sense of wonder and gratitude for the natural world, promoting practices that honor and protect it. Activities such as meditation, nature walks, and eco-therapy not only enhance personal well-being but also reinforce the idea that nurturing the Earth is a spiritual endeavor.

The integration of Gaia philosophy with spirituality also highlights the importance of community and collective action in addressing environmental challenges. Many spiritual groups advocate for environmental justice, emphasizing the need for equitable solutions that protect vulnerable communities and ecosystems. This commitment to social and environmental harmony reflects the Gaia belief that all beings are interconnected and that the health of the planet relies on the well-being of all its inhabitants.

In contemporary discussions about climate change and environmental degradation, the spiritual dimensions of Gaia philosophy provide a valuable framework for fostering deeper connections with the Earth. By embracing a spiritual approach to ecological consciousness, individuals and communities can cultivate a sense of purpose and commitment to sustainability. This alignment between spirituality and environmental stewardship encourages a holistic understanding of our place within the web of life.

Ultimately, the relationship between Gaia philosophy and spirituality invites a deeper exploration of the connections between humans and the natural world. By recognizing the Earth as a living entity deserving of reverence, we can foster a more profound commitment to protecting and nurturing our planet. This spiritual perspective not only enhances personal well-being but also contributes to the collective effort to sustain the delicate balance of life on Earth.

Pagan Beliefs and Gaia

Pagan beliefs often emphasize the reverence for nature and the interconnectedness of all living beings, making them inherently compatible with Gaia philosophy. Many pagan traditions view the Earth as a sacred entity, embodying a spirit that is worthy of respect and care. This perspective aligns closely with the principles of Gaia, which posit that the Earth functions as a self-regulating system where all life forms contribute to the overall health of the planet.

In various pagan practices, nature is not just a backdrop but a living force that interacts dynamically with human activities. Deities representing natural elements—such as earth, water, fire, and air—are central to many pagan belief systems. For instance, in Wicca, practitioners honor the Goddess as a manifestation of the Earth itself, celebrating the cycles of nature through rituals aligned with seasonal changes. This celebration of nature's rhythms reflects a deep understanding of the interconnectedness emphasized in Gaia philosophy.

Paganism often promotes the idea of animism, the belief that all entities—animals, plants, rocks, and even rivers—possess a spirit or consciousness. This view fosters a profound respect for the environment, encouraging adherents to engage with nature mindfully and to seek harmony within it. Such beliefs resonate with the Gaia perspective, which asserts that all life forms are part of a greater whole, interconnected in ways that sustain and support each other.

The emphasis on cycles and seasons in pagan traditions also aligns with Gaia philosophy's focus on ecological balance. Many pagan practices involve rituals that honor the changing seasons, acknowledging the importance of renewal and regeneration. For example, the celebration of Beltane marks the arrival of spring and the fertility of the Earth, while Samhain honors the transition into winter and the cycle of death and rebirth. These observances highlight the significance of maintaining harmony with the natural world, reflecting the self-regulating nature of ecosystems.

Moreover, contemporary pagan movements often engage actively in environmental activism, advocating for the protection of natural habitats and the preservation of biodiversity. This commitment to stewardship embodies the principles of Gaia philosophy, which emphasize humanity's responsibility to care for the Earth. Many pagans participate in initiatives aimed at combating climate change, supporting sustainable agriculture, and restoring damaged ecosystems, viewing these efforts as spiritual practices that honor the Earth.

In addition, the Gaia philosophy's emphasis on interconnectedness encourages a sense of community among pagans. Many groups engage in collective rituals that foster a shared commitment to ecological sustainability. These gatherings not only strengthen bonds among participants but also promote a collective consciousness about the importance of protecting the environment for future generations.

The relationship between pagan beliefs and Gaia philosophy ultimately reflects a shared understanding of the sacredness of nature. Both perspectives emphasize the need for a respectful and harmonious relationship with the Earth, highlighting the significance of ecological balance and sustainability. By honoring the interconnectedness of all life forms, pagans contribute to a broader dialogue about environmental ethics, emphasizing the importance of caring for the planet as a spiritual and moral imperative. This synthesis of beliefs serves as a powerful reminder of the integral role humanity plays within the Earth's self-regulating system, inspiring individuals and communities to act as stewards of the natural world.

Eastern Philosophies and Gaia

Eastern philosophies often emphasize the interconnectedness of all living beings and the importance of harmony within the natural world, principles that resonate deeply with Gaia philosophy. These traditions, rooted in ancient teachings, provide rich perspectives on the relationship between humanity and the environment, highlighting the significance of balance and respect for nature.

In Hinduism, for instance, the concept of **Prithvi**, or Earth, is revered as a goddess representing fertility and sustenance. The belief in **Ahimsa**, or non-violence, extends beyond human interactions to encompass all living beings, promoting a deep respect for the Earth and its resources. This aligns with Gaia philosophy's assertion that all life forms are interconnected and essential to the health of the planet. Rituals and practices in Hinduism often involve honoring nature, reflecting a recognition of the Earth as a living entity deserving of care and protection.

Buddhism further emphasizes the interconnectedness of life through the concept of **dependent origination**, which teaches that all phenomena arise in dependence on causes and conditions. This perspective encourages followers to recognize their place within the web of life, fostering compassion for all beings and promoting ecological mindfulness. The Buddhist practice of **mindfulness** extends to the natural world, encouraging awareness of one's actions and their impact on the environment. Such principles resonate with the idea that humanity should act as stewards of the Earth, taking responsibility for its well-being.

Taoism offers another rich perspective that aligns with Gaia philosophy. Central to Taoist thought is the concept of the **Tao**, or the Way, which emphasizes living in harmony with the natural order. This philosophy advocates for simplicity, humility, and a deep respect for nature, recognizing that all things are interconnected and that disruption to one part of the system can lead to imbalance. The practice of **Wu Wei**, or effortless action, encourages individuals to align themselves with the flow of nature rather than resist it, promoting sustainability and ecological balance.

Additionally, many indigenous philosophies across Asia, such as those found in Shintoism, view the natural world as imbued with spiritual significance. In Shinto, the belief in **Kami**, or spirits residing in natural elements such as trees, mountains, and rivers, reinforces the idea that nature is alive and sacred. This perspective fosters a deep sense of respect for the environment and emphasizes the need for harmonious coexistence with the land.

The synthesis of Eastern philosophies with Gaia principles encourages a holistic approach to environmental issues. As contemporary society grapples with challenges like climate change, biodiversity loss, and habitat destruction, the teachings from these traditions provide valuable insights for fostering sustainable practices. The emphasis on interconnectedness and the moral imperative to protect the environment can inspire individuals and communities to engage in stewardship that honors the Earth as a living organism.

Moreover, the integration of Eastern philosophies into environmental movements has led to the development of practices such as **eco-spirituality**, where ecological concerns are combined with spiritual beliefs. This movement emphasizes the need for a collective awakening to our relationship with the Earth, encouraging actions that promote ecological health and well-being.

In conclusion, the principles found in Eastern philosophies offer a profound framework for understanding humanity's role within the context of Gaia philosophy. By embracing the interconnectedness of all life and recognizing the sacredness of the natural world, individuals can cultivate a deeper commitment to protecting the planet. This alignment not only fosters a sense of responsibility for the environment but also inspires collective efforts to create a sustainable future, honoring the delicate balance that sustains life on Earth.

Gaia and Religion

The relationship between Gaia philosophy and religion presents a fascinating exploration of how various spiritual beliefs intersect with the concept of the Earth as a living, interconnected entity. Gaia philosophy emphasizes the idea that all life on Earth is interdependent and that the planet functions as a self-regulating system. This perspective resonates with numerous religious traditions that regard nature as sacred and emphasize the importance of harmony between humanity and the natural world.

Many indigenous religions possess a profound reverence for the Earth, often viewing it as a living being imbued with spirit and consciousness. For example, in many Native American cultures, the Earth is seen as "Mother Earth," a nurturing force that sustains all life. Rituals and ceremonies are frequently conducted to honor the land, water, and all living creatures, reflecting a deep understanding of the interconnectedness of life. This belief system aligns closely with Gaia philosophy, which underscores the importance of maintaining balance and respect for the environment.

In ancient Greek religion, Gaia herself was revered as the personification of the Earth, symbolizing fertility, growth, and the nurturing aspects of nature. She was seen as a foundational figure from whom all life emerged, reinforcing the idea of the Earth as a living organism. The mythology surrounding Gaia highlights the interrelationship between the divine and the natural world, showcasing how religious beliefs can shape our understanding of ecological systems.

Similarly, in Hinduism, the Earth is personified as the goddess **Bhumi**, representing fertility and the sustenance of life. Hindu teachings emphasize the sanctity of nature and the need for humans to live in harmony with the environment. Practices such as **Ahimsa**, or non-violence, extend beyond human interactions to encompass all living beings, fostering a sense of responsibility towards the Earth that mirrors the core tenets of Gaia philosophy.

Taoism also offers valuable insights that resonate with Gaia principles. Central to Taoist thought is the idea of living in harmony with the **Tao**, or the natural way of the universe. This philosophy emphasizes the interconnectedness of all things and the importance of maintaining balance within ecosystems. The practice of **Wu Wei**, or effortless action, encourages individuals to align their actions with the rhythms of nature, promoting sustainability and respect for the Earth.

Modern spiritual movements, such as eco-spirituality, integrate Gaia philosophy with religious beliefs to promote environmental awareness and stewardship. Eco-spirituality emphasizes the sacredness of the Earth and encourages individuals to recognize their role within the broader ecological system. This movement seeks to inspire a deeper connection to nature, advocating for practices that honor and protect the environment as a spiritual imperative.

Additionally, religious communities around the world are increasingly addressing environmental issues, recognizing the moral and ethical dimensions of ecological stewardship. Many faith-based organizations actively engage in conservation efforts, promote sustainable practices, and advocate for policies that protect the environment. This growing awareness reflects a broader understanding of how spiritual beliefs can inform and inspire actions to combat climate change and preserve biodiversity.

In conclusion, the intersection of Gaia philosophy and religion offers profound insights into humanity's relationship with the Earth. By recognizing the interconnectedness of all life and the sacredness of the natural world, various religious traditions provide a framework for fostering ecological awareness and responsibility. This alignment encourages individuals and communities to engage in practices that honor the Earth, promoting sustainability and stewardship as essential components of spiritual life. Through this lens, the teachings of Gaia philosophy can inspire a collective commitment to protecting the planet for future generations.

Christianity and Gaia

The relationship between Christianity and Gaia philosophy invites a thoughtful exploration of how traditional Christian beliefs intersect with the principles of interconnectedness and stewardship emphasized in Gaia thought. While Christianity historically presents a dualistic view of humanity and nature, modern interpretations increasingly acknowledge the importance of ecological responsibility and the divine presence in creation.

Central to Christianity is the belief that God created the Earth and all living beings, which lays the foundation for a spiritual relationship with the natural world. Many Christians view the Earth as God's creation, worthy of care and respect. This belief aligns with Gaia philosophy's assertion that all life forms are interconnected and that humanity has a responsibility to maintain the health of the planet.

The concept of **stewardship** is fundamental to many Christian teachings. In the Book of Genesis, humanity is tasked with the role of caretaker over creation: "Be fruitful and multiply, and fill the earth and subdue it; and have dominion over... every living thing." While this passage has often been interpreted as a call to exercise dominion, contemporary Christian thought increasingly emphasizes stewardship as responsible management and care for God's creation. This shift reflects a growing awareness of environmental issues and the necessity of preserving ecological balance, echoing the principles found in Gaia philosophy.

Furthermore, the idea of **sacredness** in creation resonates with Gaia philosophy's view of the Earth as a living organism. Many Christians emphasize that all creation reflects God's glory, suggesting that nature holds inherent value. This perspective can lead to a deeper appreciation of the interconnectedness of life, encouraging practices that honor and protect the environment. Christian environmental organizations, such as the **Evangelical Environmental Network**, promote this idea by advocating for sustainability and conservation, viewing ecological stewardship as an expression of faith.

Several Christian denominations have taken steps to address climate change and ecological degradation, integrating environmental concerns into their teachings and community practices. The Catholic Church, under the leadership of Pope Francis, has emphasized the need for ecological conversion in his encyclical, **Laudato Si'**. In this document, he calls for a collective responsibility to care for our "common home" and highlights the moral imperative to address environmental crises, reinforcing the connection between faith and ecological action.

Moreover, various Christian traditions incorporate rituals and practices that celebrate the natural world. For example, the **Feast of St. Francis**, which occurs on October 4th, includes blessings of animals and nature, emphasizing the importance of caring for all living creatures. Such practices align with Gaia philosophy's focus on the interdependence of life and the need for humans to act as guardians of the Earth.

While there are challenges in reconciling some traditional Christian doctrines with Gaia philosophy, such as differing views on humanity's role in creation, many Christians find common ground in the call to protect the environment. This intersection fosters a more holistic understanding of spirituality that embraces both faith and ecological responsibility, encouraging believers to engage actively in sustainability efforts.

In conclusion, the relationship between Christianity and Gaia philosophy illustrates a growing recognition of the importance of ecological stewardship within the Christian tradition. By embracing the interconnectedness of all life and the sacredness of creation, modern Christian interpretations can inspire believers to act responsibly toward the environment. This synergy encourages a commitment to protecting the planet, fostering a deeper understanding of humanity's role within the larger web of life and the divine presence that permeates it.

Eastern Religions and Gaia

Eastern religions offer rich insights that align closely with Gaia philosophy, emphasizing the interconnectedness of all living beings and the importance of harmony with nature. These spiritual traditions often regard the Earth as sacred, promoting a holistic understanding of life that reflects the core principles of Gaia thought.

In Hinduism, for instance, the concept of **Prithvi**, or Earth, is revered as a goddess representing fertility and sustenance. The belief in **Ahimsa**, or non-violence, extends beyond interactions among humans to encompass all living beings, fostering respect for the Earth and its resources. Rituals that honor the Earth, such as ceremonies during the harvest season, highlight the significance of maintaining a balanced relationship with nature. This worldview resonates with Gaia philosophy's assertion that all life forms are interconnected and contribute to the health of the planet.

Buddhism further emphasizes the interconnectedness of all life through the principle of **dependent origination**. This teaching illustrates that all phenomena arise in dependence on other conditions, underscoring the importance of recognizing one's place within the larger ecological system. The practice of **mindfulness** encourages an awareness of the natural world, promoting compassion for all beings and a sense of responsibility toward the environment. This alignment with Gaia philosophy fosters a commitment to living in harmony with nature, recognizing that human actions can significantly impact the delicate balance of ecosystems.

Taoism presents another perspective that complements Gaia philosophy, emphasizing the need to live in accordance with the **Tao**, or the natural way of the universe. Central to Taoist thought is the idea of **wu wei**, or effortless action, which advocates for aligning one's actions with the rhythms of nature rather than resisting them. This philosophy encourages individuals to cultivate a deep respect for the natural world and to understand that harmony is achieved by recognizing the interconnectedness of all things.

Moreover, many indigenous traditions across Asia also celebrate the sacredness of the Earth. In Shinto, for example, the belief in **Kami**—spirits residing in natural elements such as trees, rivers, and mountains—highlights the idea that nature is alive and imbued with spiritual significance. This perspective fosters a deep connection to the environment and encourages practices that honor and protect natural spaces, mirroring the principles of Gaia philosophy.

The integration of Gaia philosophy within Eastern religions has inspired contemporary spiritual movements focused on ecological awareness. Eco-spirituality combines ecological concerns with spiritual practices, encouraging individuals to recognize their connection to the Earth and engage in actions that promote ecological health. This movement emphasizes the need for collective responsibility, reflecting the belief that caring for the planet is a spiritual duty.

Furthermore, many Eastern religious communities are increasingly addressing environmental issues through advocacy and activism. Recognizing the moral and ethical dimensions of ecological stewardship, these communities promote sustainable practices and work to combat climate change. This growing awareness illustrates how spiritual teachings can inform and inspire actions to protect the environment, reinforcing the interconnectedness of all life.

In summary, Eastern religions provide valuable perspectives that resonate with Gaia philosophy, emphasizing the sacredness of nature and the interconnectedness of all living beings. By fostering a sense of responsibility and stewardship toward the Earth, these spiritual traditions encourage individuals and communities to engage actively in ecological preservation. This synergy between Eastern philosophies and Gaia principles offers a holistic framework for understanding humanity's relationship with the natural world, inspiring efforts to protect and honor the planet for future generations.

Environmental Ethics in Gaia Philosophy

Environmental ethics within the framework of Gaia philosophy emphasize the intrinsic value of all living organisms and the ecosystems they inhabit. This perspective asserts that humans are not separate from nature but rather integral components of a complex web of life. By recognizing the interconnectedness of all beings, Gaia philosophy encourages a moral responsibility to protect and sustain the planet's ecological systems.

At the heart of Gaia philosophy is the idea that Earth functions as a self-regulating entity. This notion implies that all life forms contribute to the health and balance of the planet. As such, environmental ethics derived from this philosophy promote the idea that each species, including humans, plays a vital role in maintaining the delicate equilibrium of ecosystems. This understanding challenges anthropocentric views that prioritize human interests over those of the natural world, advocating instead for a more holistic approach that respects the rights and well-being of all living beings.

One of the central tenets of environmental ethics in Gaia philosophy is the principle of **stewardship**. This concept encourages individuals and societies to act as caretakers of the Earth, recognizing their role in preserving ecological integrity for future generations. Stewardship involves making informed choices that promote sustainability, such as reducing waste, conserving resources, and protecting natural habitats. By adopting a stewardship mindset, individuals are empowered to engage in practices that foster ecological health and resilience.

Additionally, the ethics of Gaia philosophy highlight the importance of **biodiversity**. Biodiversity is crucial for the stability and resilience of ecosystems, providing essential services such as pollination, water purification, and climate regulation. Environmental ethics rooted in Gaia philosophy advocate for the protection of diverse species and habitats, emphasizing that each organism contributes to the overall functioning of the ecosystem. This perspective encourages conservation efforts aimed at preserving endangered species and restoring degraded ecosystems, recognizing that biodiversity is not merely a resource for human use but a vital component of the planet's health.

The concept of **interconnectedness** is another essential aspect of environmental ethics within Gaia philosophy. This principle highlights that actions taken in one part of the ecosystem can have far-reaching consequences. For instance, pollution in one area can affect water quality downstream, impacting both human and animal populations. Understanding these connections fosters a sense of responsibility for the broader ecological impact of individual and collective actions. It encourages a systems thinking approach, where the interdependence of life forms and their environments is acknowledged and respected.

Moreover, Gaia philosophy promotes the idea of **eco-justice**, which addresses the social and environmental injustices that arise from ecological degradation. Many marginalized communities bear the brunt of environmental harm, facing disproportionate impacts from pollution, climate change, and habitat destruction. Environmental ethics rooted in Gaia philosophy advocate for equitable solutions that consider the needs of both people and the planet. This focus on eco-justice emphasizes that protecting the environment is not only an ecological imperative but also a moral one, as it seeks to ensure that all beings have the right to a healthy and sustainable environment.

The integration of Gaia philosophy into environmental ethics also inspires collective action and grassroots movements aimed at addressing environmental challenges. Many organizations and communities are mobilizing to combat climate change, promote sustainable practices, and advocate for policies that protect the environment. By fostering a sense of community and shared responsibility, these movements align with Gaia principles, emphasizing that the health of the planet relies on the collective efforts of individuals working towards a common goal.

In conclusion, environmental ethics informed by Gaia philosophy underscore the interconnectedness of life and the moral obligation to care for the Earth. By promoting stewardship, biodiversity conservation, eco-justice, and collective action, this ethical framework encourages individuals and societies to embrace their role as caretakers of the planet. This holistic approach to environmental ethics not only honors the intricate web of life that sustains us but also inspires a deeper commitment to protecting and nurturing the Earth for generations to come.

Gaia and Sustainable Living

The principles of Gaia philosophy provide a compelling framework for sustainable living, emphasizing the interconnectedness of all life forms and the importance of maintaining ecological balance. By viewing the Earth as a self-regulating system, this philosophy encourages individuals and communities to adopt practices that promote sustainability and respect for the environment.

At the core of sustainable living is the recognition that human activities significantly impact the planet's ecosystems. Gaia philosophy highlights that every action—whether it involves resource consumption, waste production, or land use—affects the intricate web of life that sustains us. This understanding fosters a sense of responsibility, motivating individuals to make conscious choices that support ecological health.

One of the key aspects of sustainable living informed by Gaia philosophy is the promotion of **biodiversity**. Biodiversity is essential for the resilience of ecosystems, providing a variety of species that contribute to ecosystem services such as pollination, nutrient cycling, and climate regulation. Practices that encourage biodiversity, such as planting native species, creating wildlife habitats, and supporting local agriculture, align with Gaia principles by recognizing the intrinsic value of all organisms.

Another critical component of sustainable living is **resource conservation**. Gaia philosophy emphasizes the need to use resources wisely and efficiently, ensuring that future generations can also thrive. This can be achieved through practices such as reducing waste, recycling materials, and utilizing renewable energy sources. For instance, adopting solar or wind energy can help decrease reliance on fossil fuels, reducing carbon emissions and mitigating climate change. Sustainable living encourages individuals to consider the lifecycle of products, choosing options that are environmentally friendly and ethically produced.

Water conservation is also a vital aspect of sustainable living. As a finite resource, water needs to be used judiciously to ensure its availability for future generations. Practices such as rainwater harvesting, greywater recycling, and the use of water-efficient appliances can help reduce water consumption. These efforts reflect the interconnectedness emphasized in Gaia philosophy, acknowledging that water is essential not only for human survival but also for the health of ecosystems and all living beings.

Furthermore, sustainable living promotes **localism**, which involves supporting local economies and reducing the carbon footprint associated with transportation. By purchasing locally produced food and goods, individuals contribute to community resilience while minimizing the environmental impact of long-distance shipping. This practice aligns with the Gaia perspective of fostering relationships with the land and the communities that inhabit it, emphasizing the importance of local ecosystems.

Education and community engagement are also essential for promoting sustainable living practices. Raising awareness about environmental issues and encouraging collective action can inspire individuals to adopt sustainable habits. Community gardens, workshops on sustainable practices, and local clean-up events foster a sense of connection and shared responsibility, reflecting the interconnectedness central to Gaia philosophy.

Moreover, the concept of **eco-justice** plays a significant role in sustainable living. This principle emphasizes that environmental issues disproportionately affect marginalized communities, making it essential to address social equity alongside ecological sustainability. Sustainable living involves advocating for policies that protect vulnerable populations and ensuring that all individuals have access to a healthy environment. This commitment to justice aligns with Gaia philosophy by recognizing that the health of the planet and its inhabitants are inseparable.

In summary, the principles of Gaia philosophy provide a foundational framework for sustainable living that emphasizes interconnectedness, biodiversity, resource conservation, and social equity. By fostering a sense of responsibility toward the Earth and its ecosystems, individuals and communities can engage in practices that promote ecological health and resilience. This holistic approach encourages a sustainable future, honoring the delicate balance of life on our planet and ensuring that it remains vibrant for generations to come.

The Role of Balance in Ethics

In Gaia philosophy, the concept of balance plays a crucial role in shaping ethical frameworks that guide human interactions with the environment and each other. This philosophy emphasizes the interconnectedness of all life forms, suggesting that the health of the planet is maintained through a delicate equilibrium among its various components. Understanding the role of balance in ethics encourages a holistic perspective that recognizes the moral implications of our actions on the ecological web.

At the heart of this ethical perspective is the principle of **interdependence**. Gaia philosophy posits that all organisms are part of a larger ecosystem, where the well-being of one species is linked to the health of others. This interconnectedness calls for an ethical approach that considers not only human interests but also the rights and needs of other living beings. By acknowledging our role within this intricate web of life, individuals are encouraged to act with compassion and responsibility, fostering a sense of stewardship toward the environment.

The idea of balance extends to the management of natural resources as well. Sustainable practices are grounded in the understanding that overexploitation or degradation of resources can disrupt ecological balance, leading to negative consequences for both nature and humanity. Ethical considerations in resource management emphasize the need to use resources wisely and equitably, ensuring that current needs are met without compromising the ability of future generations to meet their own. This ethical approach aligns with the Gaia perspective, which advocates for harmony between human activities and the natural world.

In addition to interdependence and resource management, the notion of balance in ethics involves recognizing the **relationship between humans and nature**. Traditional ethical frameworks often place humans at the center, prioritizing human interests over environmental concerns. However, Gaia philosophy challenges this anthropocentric view by promoting an ethical stance that respects the intrinsic value of all life forms. This perspective encourages individuals to consider the broader implications of their actions on ecosystems and species, fostering a sense of humility and respect for the natural world.

Furthermore, balance in ethics is crucial for addressing social and environmental justice issues. Gaia philosophy recognizes that marginalized communities often bear the brunt of environmental degradation, facing disproportionate impacts from pollution, habitat loss, and climate change. An ethical commitment to balance involves advocating for policies that promote eco-justice, ensuring that all individuals, regardless of their socio-economic status, have access to a healthy environment. This approach emphasizes the need for equitable solutions that protect vulnerable populations while fostering a sustainable relationship with the Earth.

The role of balance in ethics also extends to the **relationship between science and spirituality**. Gaia philosophy encourages a synthesis of scientific understanding and spiritual appreciation for nature. This integration can lead to ethical frameworks that are informed by both empirical evidence and a profound respect for the Earth as a living entity. By harmonizing scientific knowledge with spiritual values, individuals can cultivate a deeper connection to the environment, fostering ethical behaviors that prioritize sustainability and ecological health.

In conclusion, the concept of balance is central to the ethical frameworks informed by Gaia philosophy. By emphasizing interconnectedness, sustainable resource management, respect for all life forms, social and environmental justice, and the integration of science and spirituality, this perspective encourages a holistic approach to ethics. Understanding the role of balance in ethics inspires individuals and communities to act with responsibility and compassion, fostering a harmonious relationship with the Earth and ensuring the well-being of all living beings.

The Concept of Gaia in Literature

The concept of Gaia has significantly influenced literature, offering a rich framework for exploring themes of interconnectedness, environmental stewardship, and the relationship between humanity and the natural world. This philosophy, which views Earth as a self-regulating organism, resonates deeply with various literary movements and genres that address ecological concerns and the intrinsic value of nature.

One of the most notable literary movements that aligns with Gaia philosophy is **Ecocriticism**. This critical approach examines the representation of nature and the environment in literature, analyzing how texts reflect and shape human attitudes toward the Earth. Ecocritics often draw on Gaia principles to emphasize the interconnectedness of life and the ethical implications of human actions on the environment. Works that highlight the relationship between characters and their natural surroundings often underscore the idea that individual and collective well-being is intertwined with the health of the planet.

Prominent authors, such as **Rachel Carson**, have effectively used literature to advocate for environmental awareness. In her groundbreaking book, **Silent Spring**, Carson illuminated the dangers of pesticide use and its devastating effects on ecosystems. Through vivid imagery and compelling narrative, she articulated the need for a harmonious relationship with nature, reflecting the core tenets of Gaia philosophy. Carson's work played a pivotal role in the modern environmental movement, demonstrating how literature can inspire awareness and action toward ecological preservation.

Contemporary writers continue to engage with the concept of Gaia in diverse ways. For instance, **Barbara Kingsolver**'s novel **Prodigal Summer** weaves together the lives of various characters, each deeply connected to the natural world around them. Through rich descriptions of the Appalachian landscape and the intricate relationships among species, Kingsolver captures the essence of Gaia philosophy, emphasizing the importance of biodiversity and ecological balance. Her narratives often celebrate the beauty of nature while also confronting the challenges posed by human activities.

Poetry, too, has served as a powerful medium for expressing Gaia principles. Poets like **Gary Snyder** and **Mary Oliver** explore themes of nature, interconnectedness, and the human spirit's relationship with the environment. Snyder's work often reflects a deep reverence for the natural world, encouraging readers to recognize their place within the larger ecological system. Similarly, Oliver's poetry captures the beauty of nature while invoking a sense of responsibility to protect it, echoing the ethical dimensions of Gaia philosophy.

In science fiction and speculative literature, the concept of Gaia has inspired narratives that explore the potential consequences of environmental degradation and climate change. Works such as **Kim Stanley Robinson**'s **Mars Trilogy** envision a future where humanity must grapple with the impact of its actions on ecosystems. Through these imaginative explorations, authors highlight the importance of understanding the Earth as a complex, interdependent system, prompting readers to consider the implications of their choices in a rapidly changing world.

Additionally, the idea of Gaia has permeated spiritual and philosophical literature, influencing discussions on the relationship between humanity and the Earth. Authors like **James Lovelock**, who coined the Gaia Hypothesis, have articulated the philosophical underpinnings of this concept, emphasizing the need for a shift in perspective regarding our relationship with nature. His writings encourage a deeper understanding of Earth's systems and inspire a sense of stewardship that aligns with Gaia principles.

The concept of Gaia also serves as a foundation for narratives that challenge anthropocentric views, inviting readers to see themselves as part of a broader ecological community. Literature that embraces Gaia philosophy often emphasizes themes of interdependence, resilience, and the necessity of living in harmony with the Earth. This approach encourages a profound reevaluation of humanity's role within the natural world, fostering a sense of connection and responsibility.

In conclusion, the concept of Gaia has significantly shaped literature by providing a framework for exploring themes of interconnectedness, environmental ethics, and the relationship between humanity and the natural world. Through various genres and movements, authors have harnessed the principles of Gaia philosophy to advocate for ecological awareness and inspire readers to reflect on their place within the web of life. This literary engagement with Gaia not only enriches our understanding of literature but also emphasizes the urgent need for a sustainable and harmonious relationship with the planet.

Depictions of Gaia Across Literary Genres

Depictions of Gaia in literature span a wide array of genres, each interpreting the philosophy's core tenets of interconnectedness, ecological balance, and the sacredness of the Earth in unique ways. From poetry to science fiction, these representations reflect humanity's evolving relationship with nature and underscore the importance of environmental stewardship.

In **poetry**, Gaia often manifests as a nurturing mother figure, representing the Earth's life-giving qualities. Poets like **Mary Oliver** and **Gary Snyder** celebrate the beauty of nature while also invoking a sense of responsibility to protect it. Oliver's work frequently emphasizes the connection between the human spirit and the natural world, encouraging readers to appreciate the wonders of flora and fauna. Snyder, influenced by Eastern philosophies and ecological awareness, often writes about the intimate bond between humans and the environment, portraying nature as a living, breathing entity deserving of reverence.

In **novels**, Gaia is depicted through characters who embody a deep connection to the Earth. For example, in **Barbara Kingsolver**'s **Prodigal Summer**, the intertwining stories of three characters illustrate the complexity of human relationships with nature. The lush descriptions of the Appalachian landscape highlight the importance of biodiversity and ecological balance, reflecting the core principles of Gaia philosophy. Kingsolver's narrative emphasizes how human actions impact the environment and advocates for a harmonious coexistence with the natural world.

Science fiction frequently explores the concept of Gaia through speculative futures that examine the consequences of environmental degradation. In **Kim Stanley Robinson**'s **Mars Trilogy**, the terraforming of Mars serves as a backdrop for discussions about humanity's responsibility to the environment. The series delves into themes of ecological restoration and sustainability, suggesting that understanding the interconnectedness of all life is essential for future survival. Robinson's work challenges readers to consider the implications of colonization and the importance of preserving ecological balance.

In **fantasy literature**, Gaia is often represented through mythical creatures and enchanted landscapes that embody the spirit of nature. In **J.R.R. Tolkien**'s **The Lord of the Rings**, the character of Ents—tree-like beings that protect forests—reflects the notion of nature as a powerful, living force. Their fight against deforestation symbolizes the struggle for ecological balance, aligning with Gaia principles. This portrayal emphasizes the need for harmony between humanity and nature, inviting readers to reflect on their relationship with the environment.

Children's literature also introduces the concept of Gaia through narratives that promote environmental awareness from a young age. Books like **The Lorax** by **Dr. Seuss** feature characters that advocate for the protection of trees and ecosystems, teaching children about the consequences of environmental neglect. These stories instill a sense of responsibility and encourage young readers to appreciate the natural world, emphasizing the interconnectedness of all living beings.

In **non-fiction**, works that explore Gaia philosophy often provide scientific insights into ecological systems and their relevance to human survival. **James Lovelock**, who coined the term "Gaia," has written extensively about the Earth as a self-regulating system. His works articulate the importance of understanding the intricate relationships between life forms and the environment, advocating for a shift in how humanity perceives its role within the ecosystem. Lovelock's writings serve as a call to action, urging individuals to adopt sustainable practices that honor the planet.

Moreover, **drama and theater** have also engaged with the concept of Gaia, using performance to evoke emotional connections to nature. Productions that incorporate environmental themes encourage audiences to reflect on their relationship with the Earth and inspire action toward sustainability. By dramatizing the consequences of environmental neglect, these performances can leave a lasting impact, resonating with the themes of Gaia philosophy.

In summary, the depictions of Gaia across various literary genres reflect a diverse array of interpretations that emphasize the interconnectedness of life and the importance of ecological balance. From poetry that celebrates the beauty of nature to speculative fiction that warns of environmental degradation, these representations highlight humanity's evolving relationship with the Earth. By engaging with the principles of Gaia philosophy, literature inspires readers to cultivate a deeper appreciation for the natural world and to act as stewards of the planet.

Gaia in Science Fiction

Science fiction often serves as a powerful medium for exploring the principles of Gaia philosophy, particularly the interconnectedness of life and the impact of human actions on the environment. This genre allows writers to imagine futures shaped by ecological realities, prompting readers to reflect on their relationship with the Earth and the consequences of their choices.

One prominent example is **Kim Stanley Robinson**'s **Mars Trilogy**, which presents a detailed vision of terraforming Mars. The series explores the complexities of creating a sustainable ecosystem on the red planet while drawing parallels to Earth's ecological challenges. Through characters who grapple with the ethical implications of their actions, Robinson delves into themes of ecological balance and the necessity of understanding the interdependence of all life forms. His portrayal of Mars as a potential new home for humanity invites readers to consider the lessons learned from Earth's environmental crises.

In **Frank Herbert**'s **Dune**, the desert planet of Arrakis serves as a crucial backdrop for discussions about resource management and ecological stewardship. The novel emphasizes the significance of water as a precious resource and explores how the planet's inhabitants adapt to their harsh environment. Herbert's work highlights the delicate balance within ecosystems and the ways in which human actions can disrupt this equilibrium. The intricate relationship between the people of Arrakis and their environment reflects Gaia philosophy's emphasis on interconnectedness and the moral responsibility to maintain ecological health.

The concept of Gaia also appears in **James Cameron**'s film **Avatar**, which depicts the lush, bioluminescent world of Pandora. The film showcases a complex ecosystem where all living beings are connected through a network known as the **Neural Tree**. This representation aligns with Gaia principles, illustrating how each species plays a vital role in maintaining the health of the environment. The conflict between the indigenous Na'vi people and human colonizers serves as a commentary on exploitation and the importance of respecting the natural world, emphasizing the consequences of disregarding ecological balance.

In the realm of speculative fiction, **Octavia Butler**'s **Parable of the Sower** explores themes of environmental collapse and societal breakdown due to climate change and resource scarcity. The narrative portrays a future where humanity grapples with the consequences of its actions, prompting readers to reflect on the importance of sustainable living. Butler's work underscores the idea that a deep understanding of interconnectedness is essential for navigating the challenges posed by environmental degradation.

Additionally, the **Gaia Hypothesis** has inspired numerous authors to envision futures where ecological balance is restored or maintained. **Octavia Butler's Kindred**, while not strictly science fiction, utilizes time travel to explore the relationships between humans and the environment, highlighting the historical contexts that shape contemporary ecological issues. The book prompts readers to consider the long-term consequences of social and environmental injustices, resonating with Gaia's themes of interconnectedness and responsibility.

Science fiction also serves as a platform for exploring radical ecological movements, as seen in **Le Guin**'s **The Dispossessed**. This novel contrasts capitalist and anarchist societies, depicting the consequences of human greed on the environment. Le Guin's work challenges readers to think critically about economic systems and their impact on ecological balance, urging a reevaluation of humanity's role within the larger ecosystem.

In summary, science fiction provides a rich landscape for examining the principles of Gaia philosophy through imaginative narratives that highlight interconnectedness, ecological balance, and the ethical implications of human actions. By envisioning futures shaped by ecological realities, authors inspire readers to reflect on their relationship with the Earth and encourage a deeper commitment to environmental stewardship. This exploration of Gaia in science fiction not only entertains but also serves as a crucial commentary on the urgent need for sustainable practices in our contemporary world.

Planet's Self-Regulation Mechanisms

In Gaia philosophy, the concept of the planet's self-regulation mechanisms plays a critical role in understanding how Earth maintains conditions that support life. This philosophy posits that living organisms interact with their environment in ways that help stabilize and regulate the Earth's systems, promoting ecological balance and resilience.

One of the key components of these self-regulating mechanisms is the feedback loops that exist within ecosystems. For example, plants play a significant role in regulating atmospheric gases through the process of photosynthesis. By absorbing carbon dioxide and releasing oxygen, plants help maintain the balance of these gases in the atmosphere. When carbon dioxide levels rise, it can lead to increased plant growth, which in turn helps to absorb more carbon dioxide, creating a positive feedback loop that contributes to climate stability.

Similarly, the role of microorganisms in soil health is crucial for self-regulation. These tiny organisms decompose organic matter, returning nutrients to the soil and making them available for plant uptake. This nutrient cycling is essential for maintaining soil fertility and ecosystem productivity. When soil health declines due to overuse or pollution, the entire ecosystem can be affected, leading to reduced plant growth and diminished capacity to regulate water and carbon cycles.

Another important self-regulating mechanism is the way ecosystems respond to disturbances. For example, forests have natural processes that promote regeneration after disturbances like wildfires. Fire can clear out underbrush and create space for new growth, which helps maintain biodiversity and ecological balance. This resilience illustrates how ecosystems can adapt and recover from disturbances, ensuring the continuation of essential ecological functions.

The oceans also exhibit remarkable self-regulation capabilities. Phytoplankton, tiny marine organisms that conduct photosynthesis, play a vital role in carbon cycling and oxygen production. By absorbing carbon dioxide from the atmosphere, they contribute significantly to mitigating climate change. Additionally, the ocean's ability to absorb heat helps regulate global temperatures, illustrating the interconnectedness of the planet's systems.

Climate regulation is another aspect of Earth's self-regulating mechanisms. The planet's climate is influenced by various factors, including ocean currents, atmospheric circulation, and the distribution of land and water. These systems work together to distribute heat and moisture, maintaining relatively stable climate conditions across different regions. However, human activities, such as deforestation and fossil fuel combustion, can disrupt these natural processes, leading to climate imbalances and exacerbating global warming.

In Gaia philosophy, the notion of Earth as a living organism underscores the importance of maintaining these self-regulating mechanisms. When humans engage in unsustainable practices, they can disrupt the delicate balance that sustains life. This highlights the ethical responsibility to protect and restore ecosystems to ensure the continued functioning of these natural processes.

Ultimately, the self-regulation mechanisms of the planet are a testament to the intricate relationships that exist within Earth's systems. By understanding these processes, individuals and societies can adopt more sustainable practices that align with the principles of Gaia philosophy. Recognizing that human well-being is deeply intertwined with the health of the planet can inspire actions that promote ecological balance and resilience, ensuring a sustainable future for all living beings.

The Earth's Feedback System

Gaia philosophy emphasizes the Earth's feedback system as a critical component of its self-regulating capabilities. This system consists of various interconnected processes that allow the planet to maintain conditions conducive to life, demonstrating how living organisms interact dynamically with their environments. Understanding these feedback mechanisms is essential for appreciating the intricate balance that sustains ecological health.

One key aspect of the Earth's feedback system is the **climate regulation** process, which involves complex interactions between the atmosphere, oceans, and land. For instance, the carbon cycle plays a vital role in regulating atmospheric carbon dioxide levels. Plants absorb carbon dioxide during photosynthesis, and when they die and decompose, the carbon stored in their biomass is returned to the atmosphere. Additionally, oceans act as significant carbon sinks, absorbing large quantities of carbon dioxide. These processes create a feedback loop: as carbon dioxide levels rise due to human activities, plants and oceans work to absorb this excess, thereby helping to stabilize the climate.

Another critical feedback mechanism is the **water cycle**, which regulates moisture and temperature across the planet. Evaporation from bodies of water leads to cloud formation, resulting in precipitation that nourishes ecosystems. When climate conditions change, such as during periods of drought, reduced precipitation can lead to water scarcity, affecting plant growth and, consequently, the entire food web. Conversely, healthy ecosystems can help maintain local climates by moderating temperatures and increasing humidity through transpiration. This interconnectedness highlights how changes in one part of the system can influence the entire feedback mechanism.

The role of **biodiversity** within the Earth's feedback system is also paramount. Diverse ecosystems are more resilient to disturbances, such as climate fluctuations, pests, and diseases. For instance, a diverse range of plant species can contribute to soil stability, nutrient cycling, and habitat diversity, which in turn supports a variety of animal life. When biodiversity is diminished, as seen through habitat destruction or species extinction, the resilience of ecosystems is compromised, leading to further instability within the feedback system. This interplay underscores the importance of maintaining biodiversity to ensure the robustness of ecological processes.

Geological processes contribute to the Earth's feedback system as well. For example, volcanic eruptions release gases, including carbon dioxide and sulfur dioxide, into the atmosphere. While volcanic activity can temporarily affect climate conditions, the long-term effects of these gases are moderated by the carbon cycle, as natural processes eventually restore balance. Such interactions demonstrate how geological and biological processes work together to regulate the planet's environment.

Human activities have significantly impacted these feedback systems, often leading to unintended consequences. Deforestation, pollution, and the burning of fossil fuels increase greenhouse gas emissions, disrupting the natural carbon cycle and altering climate patterns. These actions can trigger feedback loops that exacerbate climate change, such as the melting of polar ice, which reduces the Earth's albedo effect (its ability to reflect sunlight), leading to further warming. Recognizing these impacts emphasizes the ethical responsibility to engage in sustainable practices that honor the Earth's feedback mechanisms.

In summary, the Earth's feedback system is a complex network of processes that illustrate the principles of Gaia philosophy, emphasizing the interconnectedness of life and the environment. By understanding these feedback mechanisms, individuals and communities can better appreciate the importance of ecological balance and the need for responsible stewardship of the planet. This awareness fosters a commitment to preserving the Earth's systems, ensuring a sustainable future for all living organisms.

The Role of Biodiversity

Biodiversity plays a crucial role in Gaia philosophy, underpinning the concept that the Earth functions as a self-regulating system where all living organisms are interconnected. The variety of life forms—ranging from microorganisms to plants and animals—contributes significantly to the stability and resilience of ecosystems, reinforcing the idea that each species plays a vital role in maintaining the health of the planet.

One of the primary functions of biodiversity is its contribution to **ecosystem stability**. Diverse ecosystems are better equipped to withstand disturbances such as climate fluctuations, natural disasters, and disease outbreaks. For example, a forest with a wide range of tree species is more resilient to pests and diseases compared to a monoculture forest. This diversity allows different species to fill various ecological niches, ensuring that essential functions—such as pollination, nutrient cycling, and soil formation—continue even when certain species are impacted by stressors.

Biodiversity also enhances **ecosystem services**, which are the benefits that humans derive from nature. These services include provisioning (such as food, water, and raw materials), regulating (such as climate regulation and flood control), supporting (such as nutrient cycling and soil formation), and cultural services (such as recreational opportunities and spiritual benefits). A diverse range of species contributes to these services in unique ways, ensuring that ecosystems can provide for human needs while maintaining their health. For example, wetlands with diverse plant species are more effective at filtering pollutants and regulating water flow, thereby improving water quality and reducing flood risks.

The interconnectedness emphasized in Gaia philosophy highlights that biodiversity is not merely a collection of species but a complex web of relationships. Each organism, from the smallest bacteria to the largest mammals, plays a role in the ecosystem's functioning. For instance, pollinators like bees and butterflies are essential for the reproduction of many flowering plants, which, in turn, provide food and habitat for other species. Disrupting this relationship by harming pollinator populations can lead to declines in plant diversity and the animals that depend on those plants, showcasing how the loss of one species can have cascading effects throughout the ecosystem.

In addition to ecological functions, biodiversity is critical for **adaptation and resilience** in the face of environmental changes. A diverse genetic pool within species allows populations to adapt to shifting conditions, such as changing climates or new diseases.

For example, a population of plants with a wide variety of genetic traits is more likely to produce individuals that can survive in altered environments compared to a genetically uniform population. This adaptability is essential for the long-term survival of ecosystems, reinforcing the Gaia philosophy's emphasis on the importance of balance and interdependence.

However, biodiversity is currently under threat due to human activities such as habitat destruction, pollution, climate change, and overexploitation. The decline in species diversity can lead to the disruption of ecosystem functions and services, ultimately compromising the health of the planet. Gaia philosophy underscores the ethical responsibility to protect biodiversity, recognizing that the well-being of humanity is intrinsically linked to the health of the Earth's ecosystems.

Efforts to conserve biodiversity align with the principles of Gaia philosophy by promoting sustainable practices that honor the interconnectedness of life. Initiatives such as habitat restoration, wildlife protection, and the establishment of protected areas aim to preserve ecosystems and the diverse species within them. By recognizing the importance of biodiversity in maintaining ecological balance, individuals and communities can engage in actions that support the health of the planet.

In conclusion, biodiversity is fundamental to the principles of Gaia philosophy, emphasizing the interconnectedness and interdependence of all living organisms. Its role in maintaining ecosystem stability, providing essential services, and fostering resilience underscores the ethical imperative to protect the planet's diverse life forms. By valuing and conserving biodiversity, humanity can work toward a sustainable future that honors the delicate balance of life on Earth.

Environmentalism and Gaia Philosophy

Environmentalism and Gaia philosophy share a deep commitment to understanding and preserving the intricate relationships between living organisms and their environments. At the heart of Gaia philosophy is the idea that Earth functions as a self-regulating system, where the interdependence of all life forms contributes to the stability and health of the planet. This holistic perspective aligns closely with the principles of environmentalism, which advocates for the protection of ecosystems and the promotion of sustainable practices.

One of the central tenets of Gaia philosophy is the recognition of the interconnectedness of life. Every species, no matter how small, plays a vital role in maintaining ecological balance. This understanding fosters a sense of responsibility among individuals and communities to protect natural habitats and biodiversity. Environmentalism, in turn, emphasizes the importance of conserving ecosystems and safeguarding the myriad species that inhabit them. By highlighting the consequences of human actions on the environment, both movements advocate for a more ethical and sustainable relationship with nature.

Gaia philosophy also emphasizes the role of feedback mechanisms in maintaining the Earth's systems. These feedback loops demonstrate how changes in one part of the ecosystem can affect the entire system. For example, the decline of a keystone species can lead to significant changes in community structure and ecosystem function. Environmentalists utilize this understanding to promote policies and practices that consider the broader implications of ecological disruption, reinforcing the need for proactive conservation efforts.

The integration of Gaia philosophy into environmentalism encourages a systems-thinking approach, where the focus is not solely on individual species or resources but on the complex interactions that sustain life. This perspective is particularly relevant in addressing pressing global issues such as climate change, habitat destruction, and pollution. By recognizing the interdependence of ecosystems, environmentalists can advocate for comprehensive solutions that tackle the root causes of environmental degradation rather than merely addressing symptoms.

Another significant aspect of this relationship is the ethical imperative to consider the rights of non-human beings. Gaia philosophy posits that all life forms possess intrinsic value, challenging anthropocentric views that prioritize human interests above those of the natural world. Environmentalism often incorporates this ethical framework, advocating for policies that protect endangered species and preserve natural habitats. This approach not only fosters respect for biodiversity but also emphasizes the moral responsibility to safeguard the planet for future generations.

Moreover, the Gaia perspective encourages a shift in societal values towards sustainability. It promotes the idea that human well-being is intricately connected to the health of the planet. This understanding can inspire individuals to adopt more sustainable lifestyles, from reducing waste and conserving resources to supporting local and organic agriculture. Environmentalism often builds on this ethos, emphasizing the importance of collective action in fostering a sustainable future.

The intersection of Gaia philosophy and environmentalism is also reflected in the growing movement toward **eco-spirituality**. This movement seeks to integrate ecological awareness with spiritual practices, emphasizing the sacredness of the Earth and the need for stewardship. Eco-spirituality encourages individuals to connect deeply with nature, recognizing that caring for the environment is not just a physical responsibility but also a spiritual one. This approach resonates with the principles of Gaia philosophy, reinforcing the idea that the Earth is a living entity deserving of reverence.

In conclusion, the relationship between environmentalism and Gaia philosophy is marked by a shared commitment to understanding and preserving the interconnectedness of life on Earth. By advocating for ethical stewardship, recognizing the importance of biodiversity, and promoting sustainable practices, both movements work towards creating a healthier planet. The principles of Gaia philosophy enrich environmentalism by providing a holistic framework that emphasizes the need for balance, respect, and responsibility in our interactions with the natural world. This synergy inspires individuals and communities to engage in meaningful actions that honor the delicate web of life that sustains us all.

The Green Movement and Gaia

The Green Movement embodies a collective response to environmental degradation, advocating for sustainability and ecological responsibility. This movement aligns closely with Gaia philosophy, which posits that Earth functions as a self-regulating system, where the interconnectedness of all living organisms plays a crucial role in maintaining ecological balance. By promoting awareness of our relationship with the environment, the Green Movement draws on Gaia principles to inspire action toward a more sustainable future.

At the core of the Green Movement is the recognition that human activities have a significant impact on the planet. From pollution and deforestation to climate change and habitat destruction, the consequences of industrialization and urbanization threaten the delicate balance that sustains life. Gaia philosophy emphasizes that all life forms, including humans, are interconnected and that the health of one depends on the health of the whole. This perspective encourages activists and individuals to consider the broader implications of their actions, fostering a sense of responsibility for the Earth's well-being.

One of the movement's key focuses is promoting **biodiversity conservation**. The variety of life on Earth is essential for ecosystem stability, providing numerous services that humans rely on, such as clean air, water, and food. By protecting natural habitats and advocating for sustainable practices, the Green Movement aligns with Gaia philosophy's assertion that every species has intrinsic value and contributes to the overall health of the planet. Conservation efforts often emphasize restoring ecosystems and protecting endangered species, reinforcing the idea that preserving biodiversity is crucial for maintaining ecological balance.

The Green Movement also emphasizes **sustainable practices** that minimize environmental impact. This includes promoting renewable energy sources, such as solar and wind power, which help reduce reliance on fossil fuels and mitigate climate change. The shift towards sustainable agriculture, which prioritizes organic farming and permaculture, echoes Gaia principles by encouraging practices that work in harmony with natural processes rather than exploiting them. Such approaches highlight the importance of living sustainably within the planet's ecological limits, reflecting the interconnectedness emphasized in Gaia philosophy.

Another critical aspect of the Green Movement is its advocacy for **environmental justice**. Many marginalized communities disproportionately bear the brunt of environmental degradation, facing challenges such as pollution, lack of access to clean water, and vulnerability to climate change impacts. The Green Movement seeks to address these inequalities by advocating for policies that promote fair distribution of environmental benefits and burdens. This commitment to social equity resonates with Gaia philosophy's ethical dimensions, reinforcing the idea that a healthy planet is essential for the well-being of all its inhabitants.

Education and grassroots activism are also fundamental components of the Green Movement. By raising awareness about environmental issues and engaging communities in conservation efforts, activists encourage individuals to adopt more sustainable lifestyles. Educational initiatives often incorporate the principles of Gaia philosophy, promoting a sense of connection to the natural world and fostering an understanding of how individual actions can contribute to collective ecological health.

The rise of **eco-spirituality** within the Green Movement further illustrates the influence of Gaia philosophy. Many individuals and groups blend environmental activism with spiritual practices, emphasizing the sacredness of the Earth and the need for stewardship. This approach inspires a deeper emotional connection to nature, encouraging people to see themselves as part of a larger ecological community. By fostering a sense of reverence for the planet, eco-spirituality aligns with Gaia principles and motivates individuals to take meaningful action to protect the environment.

In summary, the Green Movement is deeply intertwined with Gaia philosophy, emphasizing the interconnectedness of all life forms and the importance of maintaining ecological balance. By advocating for biodiversity conservation, sustainable practices, environmental justice, education, and eco-spirituality, the movement encourages a holistic approach to environmental stewardship. This synergy inspires individuals and communities to recognize their role in protecting the planet, fostering a sustainable future that honors the delicate web of life that sustains us all.

Ecological Activism Inspired by Gaia

Ecological activism inspired by Gaia philosophy emphasizes the interconnectedness of all living beings and the importance of maintaining ecological balance. This philosophy provides a framework for understanding the Earth as a self-regulating organism, where human actions are deeply intertwined with the health of the planet. Activists draw on these principles to advocate for environmental protection, sustainability, and social justice, motivating collective action toward a healthier world.

One of the key inspirations for ecological activism from Gaia philosophy is the emphasis on **interdependence**. Activists recognize that every species plays a role in sustaining the ecosystem and that the decline of one can have cascading effects on others. This understanding drives campaigns aimed at protecting biodiversity, as activists work to safeguard habitats and restore ecosystems that support diverse life forms. Initiatives like reforestation projects, wetland restoration, and wildlife protection efforts illustrate how the principles of Gaia can inform practical actions to preserve the intricate web of life.

The concept of **feedback mechanisms** within Gaia philosophy also inspires activists to address the consequences of human actions on the environment. For example, climate change represents a significant disruption to the planet's self-regulating systems. Activists raise awareness about the impact of carbon emissions, deforestation, and pollution on global climate patterns, advocating for policies that promote renewable energy, reduce waste, and mitigate environmental harm. By highlighting these feedback loops, they encourage a holistic understanding of ecological issues, emphasizing that every action has far-reaching consequences.

Grassroots movements often embody the principles of Gaia philosophy by emphasizing community engagement and local solutions to environmental challenges. Many activists work to empower communities to take action on issues such as pollution, habitat destruction, and food insecurity. Community gardens, clean-up campaigns, and local conservation initiatives foster a sense of ownership and responsibility for the environment. This grassroots approach aligns with the Gaia philosophy's emphasis on interconnectedness, as it highlights the importance of local ecosystems and the role individuals play in sustaining them.

The integration of **eco-justice** into ecological activism is another significant aspect inspired by Gaia philosophy. Activists recognize that environmental degradation often disproportionately affects marginalized communities, exacerbating social inequalities.

This awareness drives efforts to address both environmental and social justice issues, advocating for equitable policies that protect vulnerable populations and promote access to clean air, water, and green spaces. By emphasizing that a healthy environment is a fundamental right for all, activists work toward systemic change that honors the interconnectedness of human and ecological well-being.

Education and awareness-raising are crucial components of ecological activism influenced by Gaia philosophy. Many activists focus on educating the public about the principles of sustainability, biodiversity, and ecological stewardship. Workshops, community events, and social media campaigns serve to inform and inspire individuals to adopt more environmentally friendly practices. By fostering a deeper understanding of the interconnectedness of life, these educational initiatives empower people to take action in their own lives and communities.

The role of **art and culture** in ecological activism also reflects the influence of Gaia philosophy. Artists and writers often draw upon the themes of interconnectedness and environmental stewardship to create works that resonate emotionally with audiences. From visual art that celebrates nature to literature that explores humanity's relationship with the Earth, creative expressions can inspire individuals to reflect on their role within the ecosystem. These cultural narratives can galvanize support for environmental causes and promote a sense of collective responsibility toward the planet.

In summary, ecological activism inspired by Gaia philosophy emphasizes the interconnectedness of life and the importance of maintaining ecological balance. By advocating for biodiversity conservation, addressing climate change, promoting eco-justice, fostering community engagement, and integrating education and culture into their efforts, activists draw upon the principles of Gaia to inspire meaningful action. This approach not only highlights the intricate relationships within ecosystems but also reinforces the moral imperative to protect the planet for current and future generations.

Implementing Gaia Philosophy in Everyday Life

Implementing the principles of Gaia philosophy in everyday life involves recognizing the interconnectedness of all living beings and the importance of maintaining ecological balance. By embracing this philosophy, individuals can make conscious choices that contribute to the health of the planet and foster a more sustainable future.

One of the fundamental ways to incorporate Gaia philosophy into daily life is through **sustainable living practices**. This includes reducing waste, conserving resources, and choosing products that have minimal environmental impact. Simple actions, such as using reusable bags, bottles, and containers, can significantly reduce plastic waste. Composting organic materials and recycling can also minimize the strain on landfills and promote a circular economy. By adopting these practices, individuals can help maintain the Earth's resources and support its self-regulating systems.

In the realm of **food choices**, embracing Gaia principles means opting for locally sourced and organic produce. Supporting local farmers and businesses not only reduces the carbon footprint associated with transportation but also fosters a connection to the community and the environment. By choosing seasonal foods, individuals can align their diets with natural cycles, promoting biodiversity and sustainable agriculture. Additionally, growing one's own food, whether through gardening or participating in community gardens, allows for a direct engagement with nature and a deeper appreciation for the ecosystems that support life.

Engaging with **nature** is another essential aspect of implementing Gaia philosophy. Spending time outdoors—whether hiking, walking in a park, or simply enjoying a garden—fosters a sense of connection to the Earth. This engagement can inspire individuals to develop a deeper appreciation for the natural world and the importance of preserving it. Practicing mindfulness in nature, such as through eco-therapy or nature meditation, enhances well-being while reinforcing the understanding that humans are part of a larger ecological system.

Advocacy and community involvement are also vital for implementing Gaia philosophy in daily life. Joining local environmental organizations or participating in community clean-up events can amplify individual efforts and promote collective action toward ecological sustainability. Engaging in discussions about environmental issues and supporting policies that protect natural habitats and promote sustainability can create a ripple effect, encouraging others to adopt similar values and practices.

Education plays a crucial role in the practical application of Gaia principles. Learning about ecological systems, sustainability practices, and environmental challenges equips individuals with the knowledge necessary to make informed choices. This can involve reading books, attending workshops, or participating in online courses related to ecology and sustainability. Sharing this knowledge with others can help foster a community committed to environmental stewardship.

Incorporating **eco-friendly habits** into everyday routines can also reflect Gaia philosophy. This may include reducing energy consumption by using energy-efficient appliances, turning off lights when not in use, or using public transportation and carpooling to decrease carbon emissions. Small changes, such as switching to LED light bulbs or utilizing renewable energy sources like solar power, contribute to a larger collective effort toward a sustainable future.

Finally, cultivating an **ethical mindset** that values the intrinsic worth of all living beings is fundamental to implementing Gaia philosophy. This includes recognizing the rights of non-human animals and promoting compassion in all interactions. Supporting cruelty-free products, adopting a plant-based diet, or advocating for animal rights can align personal values with the interconnectedness emphasized in Gaia philosophy.

In summary, implementing Gaia philosophy in everyday life involves adopting sustainable practices, engaging with nature, participating in community efforts, educating oneself and others, and cultivating an ethical mindset. By recognizing the interconnectedness of all life forms and the importance of ecological balance, individuals can contribute to the health of the planet and foster a sustainable future. Embracing these principles not only enhances personal well-being but also strengthens the collective effort to protect the Earth for generations to come.

Eco-friendly Choices and Behaviours

Making eco-friendly choices and adopting sustainable behaviors are fundamental aspects of Gaia philosophy, which emphasizes the interconnectedness of all living beings and the importance of maintaining ecological balance. By integrating environmentally conscious practices into daily life, individuals can contribute to the health of the planet and support its self-regulating systems.

One of the most impactful ways to make eco-friendly choices is through **reducing waste**. Simple actions such as using reusable bags, water bottles, and containers can significantly decrease the amount of plastic waste generated. Composting organic materials not only reduces landfill waste but also enriches soil health, aligning with Gaia principles by promoting nutrient cycling within ecosystems. Implementing a "zero waste" lifestyle encourages individuals to think critically about their consumption habits, fostering a deeper awareness of the impact their choices have on the environment.

Sustainable food choices are another crucial area for promoting eco-friendly behaviors. Opting for locally sourced and seasonal produce reduces the carbon footprint associated with transporting food over long distances. Supporting organic farming practices helps minimize the use of synthetic pesticides and fertilizers, which can harm local ecosystems. Additionally, adopting a plant-based diet or reducing meat consumption can significantly lower an individual's environmental impact, as livestock farming is resource-intensive and contributes to greenhouse gas emissions.

Integrating **energy efficiency** into daily routines is also essential for sustainable living. Simple measures, such as using energy-efficient appliances, turning off lights when not in use, and utilizing natural light during the day, can help reduce energy consumption. Transitioning to renewable energy sources, such as solar or wind power, further supports ecological sustainability. By choosing to invest in green energy solutions, individuals contribute to the reduction of fossil fuel reliance and the associated environmental degradation.

Water conservation is another critical aspect of eco-friendly behavior. Implementing practices like fixing leaks, using low-flow fixtures, and collecting rainwater for gardening can help preserve this vital resource. Understanding the role of water in sustaining ecosystems reinforces the interconnectedness emphasized in Gaia philosophy, highlighting that responsible water use is essential for maintaining ecological balance.

Engaging in **sustainable transportation** options can significantly reduce an individual's carbon footprint. Walking, biking, carpooling, or using public transportation not only minimizes greenhouse gas emissions but also promotes a healthier lifestyle. For those who need to drive, choosing fuel-efficient or electric vehicles can further decrease environmental impact. Community initiatives that promote shared transportation solutions can foster a culture of sustainability, encouraging collective responsibility.

Practicing **mindfulness in consumption** also reflects Gaia philosophy's principles. This involves making conscious choices about products and services, prioritizing those that are eco-friendly, ethically sourced, and produced with minimal environmental impact. Reading labels, supporting brands with sustainable practices, and considering the lifecycle of products can empower individuals to make informed decisions that align with their values.

Educating oneself and others about environmental issues is a vital component of fostering eco-friendly behaviors. Engaging with local environmental groups, attending workshops, or participating in community events can raise awareness about sustainability and inspire collective action. Sharing knowledge and experiences within social circles encourages a culture of environmental stewardship, reinforcing the interconnectedness of individual and community actions.

Finally, embracing an **ethics of care** for the planet involves recognizing the intrinsic value of all living beings. This perspective promotes compassion in interactions with nature, leading to choices that respect the rights of animals and ecosystems. Supporting cruelty-free products, advocating for wildlife conservation, and participating in local restoration projects can help nurture this ethical commitment.

In conclusion, making eco-friendly choices and adopting sustainable behaviors are essential for embodying the principles of Gaia philosophy. By recognizing the interconnectedness of all life and the importance of maintaining ecological balance, individuals can contribute to the health of the planet through conscious actions. Embracing these practices not only supports environmental sustainability but also fosters a deeper connection to the natural world, ensuring that future generations can thrive in a healthy and vibrant ecosystem.

Gaia-inspired Community Living

Gaia-inspired community living emphasizes the principles of interconnectedness, sustainability, and collective responsibility toward the environment. This approach fosters a sense of belonging and collaboration among individuals while promoting practices that support ecological health. By prioritizing community engagement and sustainable practices, Gaia-inspired living creates vibrant, resilient environments that reflect the philosophy's core tenets.

One fundamental aspect of this lifestyle is the creation of **eco-communities**, which are designed to integrate sustainable practices into daily life. These communities often prioritize green building techniques, such as using sustainable materials, energy-efficient designs, and incorporating renewable energy sources. By designing homes and communal spaces that minimize environmental impact, eco-communities reflect the belief that human habitation can coexist harmoniously with nature.

In addition to sustainable architecture, these communities typically emphasize **shared resources** to reduce consumption and waste. Community gardens, tool libraries, and shared transportation systems foster a culture of collaboration and resourcefulness. By pooling resources, community members can reduce their individual environmental footprints while cultivating a strong sense of community. Gardening together not only provides fresh produce but also creates opportunities for education and connection, allowing members to engage with nature and each other.

Social equity is another vital principle in Gaia-inspired community living. Such communities often strive to create inclusive environments that support diverse populations. This can involve establishing affordable housing, accessible public spaces, and community programs that promote social engagement and well-being. By prioritizing inclusivity and collaboration, these communities reflect the interconnectedness emphasized in Gaia philosophy, recognizing that the health of the community is intrinsically linked to the health of the environment.

Education plays a critical role in fostering Gaia-inspired community living. Many communities implement educational programs that promote awareness of ecological issues, sustainability practices, and the importance of biodiversity. Workshops, seminars, and hands-on experiences encourage individuals to adopt eco-friendly behaviors and understand their impact on the environment. This emphasis on education empowers community members to take an active role in environmental stewardship and fosters a sense of collective responsibility.

Local economies are often prioritized in Gaia-inspired communities, emphasizing the importance of supporting local businesses and reducing reliance on global supply chains. By promoting farmers' markets, community-supported agriculture (CSA), and local artisans, these communities create economic systems that reflect ecological values. Supporting local economies not only strengthens community ties but also reduces transportation emissions and promotes sustainable practices.

Participatory governance is another characteristic of Gaia-inspired community living. Many eco-communities adopt decision-making processes that involve all members, ensuring that everyone has a voice in shaping their environment. This democratic approach fosters transparency and accountability, allowing communities to respond collectively to challenges and adapt to changing circumstances. Engaging in community governance aligns with the Gaia philosophy's emphasis on interconnectedness, as it recognizes that individual actions contribute to the collective well-being.

Environmental restoration and conservation efforts are often integral to Gaia-inspired communities. Members may engage in projects aimed at restoring local habitats, such as reforestation, wetland restoration, or invasive species removal. By actively participating in ecological restoration, community members deepen their connection to the land and contribute to the overall health of their environment.

In summary, Gaia-inspired community living embodies the principles of interconnectedness, sustainability, and collective responsibility. By fostering eco-communities that prioritize shared resources, social equity, education, local economies, participatory governance, and environmental restoration, individuals can create vibrant, resilient environments that reflect Gaia philosophy. This approach not only enhances the quality of life for community members but also contributes to the health of the planet, ensuring a sustainable future for generations to come.

Gaia and the Future

The principles of Gaia philosophy offer a compelling framework for envisioning a sustainable future, emphasizing the interconnectedness of all living beings and the importance of maintaining ecological balance. As humanity faces pressing environmental challenges, including climate change, biodiversity loss, and resource depletion, the insights provided by Gaia philosophy can guide collective action and inform policies that promote a healthier planet.

One of the key aspects of Gaia philosophy is the recognition that Earth operates as a self-regulating system. This understanding underscores the importance of preserving biodiversity, as each species contributes to the overall health and stability of ecosystems. Moving forward, a commitment to conservation and the protection of natural habitats will be essential for maintaining this balance. Efforts such as reforestation, wetland restoration, and the establishment of protected areas can help restore ecosystems and enhance their resilience in the face of environmental stressors.

In addressing climate change, Gaia philosophy advocates for a holistic approach that considers the interconnections among various ecological systems. This perspective encourages the adoption of sustainable practices across sectors, including energy, transportation, and agriculture. Transitioning to renewable energy sources, such as solar and wind, will be crucial for reducing greenhouse gas emissions and mitigating climate impacts. Additionally, promoting sustainable agricultural practices, like regenerative farming and agroecology, can enhance soil health, increase biodiversity, and improve food security.

The emphasis on **localism** within Gaia philosophy highlights the importance of community engagement and grassroots movements in creating a sustainable future. Localized food systems, such as farmers' markets and community-supported agriculture, can reduce transportation emissions and promote healthier diets. By empowering communities to take an active role in shaping their environments, we can foster a sense of stewardship and collective responsibility that aligns with Gaia principles.

Education will play a critical role in advancing Gaia philosophy in the future. By fostering environmental literacy and awareness of ecological issues, we can inspire individuals to adopt more sustainable lifestyles and advocate for systemic change. Educational initiatives that incorporate hands-on experiences, such as community gardens and nature-based learning, can deepen the connection between individuals and the natural world. This engagement will cultivate a generation of environmentally conscious citizens who are committed to preserving the planet.

The concept of **eco-justice** is also central to the future envisioned by Gaia philosophy. As environmental degradation disproportionately affects marginalized communities, it is crucial to advocate for equitable solutions that address social and environmental inequalities. Ensuring that all individuals have access to clean air, water, and green spaces is essential for fostering a sustainable and just society. This commitment to eco-justice reinforces the interconnectedness of social and ecological systems, highlighting that the health of the planet and the well-being of its inhabitants are inseparable.

Moreover, fostering a sense of wonder and appreciation for the natural world will be vital for creating a sustainable future. Artistic expressions, literature, and cultural narratives that celebrate the beauty of nature can inspire individuals to connect with their environment on a deeper level. By recognizing the intrinsic value of all living beings, we can cultivate a collective ethic that prioritizes the health of the Earth and encourages proactive stewardship.

As humanity moves forward, the integration of Gaia philosophy into decision-making processes at local, national, and global levels will be essential for addressing the complex challenges facing the planet. Policymakers must consider the interconnectedness of ecological systems and the long-term implications of their actions. Collaborative approaches that involve multiple stakeholders, including governments, businesses, and communities, can lead to innovative solutions that promote sustainability and resilience.

In summary, the future shaped by Gaia philosophy emphasizes the interconnectedness of life, the importance of biodiversity, and the need for sustainable practices. By fostering community engagement, promoting environmental education, advocating for eco-justice, and inspiring a sense of wonder for the natural world, we can work toward a healthier planet. Embracing these principles will not only enhance the quality of life for current generations but also ensure that future generations inherit a vibrant and sustainable Earth.

Future Predictions Based on Gaia Philosophy

Future predictions based on Gaia philosophy emphasize the intricate interconnectedness of all life forms and the essential balance within ecosystems. This perspective provides valuable insights into potential ecological outcomes and the importance of proactive stewardship in addressing environmental challenges. As humanity navigates an era marked by climate change, biodiversity loss, and resource depletion, Gaia philosophy encourages a forward-thinking approach that recognizes the profound implications of human actions on the planet's health.

One significant prediction is the increasing importance of **biodiversity conservation**. As ecosystems face mounting pressures from habitat destruction, climate change, and pollution, the loss of biodiversity could lead to irreversible consequences for ecological balance. Future efforts will likely focus on protecting and restoring habitats, preserving endangered species, and promoting sustainable land-use practices. Recognizing that each species plays a vital role in maintaining ecosystem stability, conservation initiatives will aim to enhance resilience and adaptability in the face of environmental stressors.

Gaia philosophy also suggests a shift toward **sustainable resource management** as a crucial element of future predictions. With the global population projected to reach nearly 10 billion by 2050, the demand for resources will continue to escalate. This will necessitate a fundamental change in how societies approach consumption and production. Transitioning to circular economies, where resources are reused, recycled, and repurposed, will be essential in minimizing waste and conserving natural resources. Innovative practices such as urban agriculture, renewable energy systems, and green infrastructure will play a vital role in creating sustainable communities.

Another key prediction is the growing recognition of the need for **holistic approaches** to environmental issues. Future policymaking will likely emphasize integrated solutions that consider the complex interactions between social, economic, and ecological systems. By adopting a systems-thinking perspective, policymakers can develop strategies that address multiple challenges simultaneously, such as climate change mitigation, poverty alleviation, and public health improvement. Collaborative efforts involving diverse stakeholders—governments, businesses, NGOs, and communities—will be crucial in driving meaningful change.

The concept of **eco-justice** will also become increasingly prominent in future environmental discussions. As climate change disproportionately impacts marginalized communities, there will be a heightened focus on ensuring equitable access to resources and opportunities for all individuals. Future initiatives will likely prioritize social equity, recognizing that environmental sustainability cannot be achieved without addressing the systemic inequalities that exist within society. Advocacy for policies that promote environmental justice will become essential in building resilient and inclusive communities.

Furthermore, advancements in **technology** will significantly shape the future as they relate to Gaia philosophy. Innovations in renewable energy, sustainable agriculture, and conservation technology will provide new tools for promoting ecological balance. For instance, precision agriculture techniques can optimize resource use while minimizing environmental impact, helping to ensure food security in an era of climate uncertainty. Additionally, data-driven approaches to monitoring ecosystems will enhance our understanding of ecological dynamics, enabling more effective conservation efforts.

The role of **education and awareness** will also be crucial in shaping future predictions based on Gaia philosophy. As individuals become more informed about environmental issues, there will likely be a growing demand for sustainable practices in daily life. Educational initiatives that emphasize the interconnectedness of life and the importance of ecological stewardship will foster a culture of responsibility and action. By instilling values of sustainability in future generations, societies can cultivate a sense of collective responsibility for the planet's health.

Finally, the future may see an increased emphasis on **spiritual and ethical dimensions** of environmentalism influenced by Gaia philosophy. Many people are recognizing the intrinsic value of nature and the moral imperative to protect it. This shift in perspective may lead to a deeper appreciation for the sacredness of the Earth, inspiring individuals and communities to engage in eco-spiritual practices that honor the interconnectedness of all life. By blending ecological awareness with spiritual values, a holistic approach to environmental stewardship can emerge, further reinforcing the principles of Gaia.

In conclusion, future predictions grounded in Gaia philosophy emphasize the importance of biodiversity conservation, sustainable resource management, holistic approaches to environmental challenges, eco-justice, technological advancements, education, and spiritual dimensions of ecological stewardship. By recognizing the interconnectedness of all life and the need for balance within ecosystems, individuals and communities can work together to create a sustainable and resilient future for the planet. Embracing these principles will ensure that we honor the delicate web of life that sustains us all and foster a healthier environment for generations to come.

Gaia and Futuristic Technologies

Futuristic technologies hold great potential for advancing the principles of Gaia philosophy, which emphasizes the interconnectedness of all life and the importance of maintaining ecological balance. As humanity faces pressing environmental challenges, innovative technologies can play a pivotal role in fostering sustainability and promoting harmonious relationships with the planet.

One of the most promising areas of development is in **renewable energy technologies**. Solar panels, wind turbines, and advanced energy storage systems are becoming increasingly efficient and affordable, enabling a transition away from fossil fuels. By harnessing the power of natural resources, these technologies align with Gaia philosophy by reducing carbon emissions and minimizing the ecological footprint of energy consumption. Innovations such as floating solar farms and vertical wind turbines also demonstrate how energy solutions can be integrated into existing ecosystems without causing significant disruption.

Smart grid technology is another significant advancement that supports Gaia principles. By optimizing the distribution of electricity through real-time data and communication, smart grids can enhance energy efficiency and reliability. This technology enables the integration of renewable energy sources, allowing for a more resilient and sustainable energy system. By reducing waste and improving energy management, smart grids reflect the interconnectedness emphasized in Gaia philosophy, ensuring that energy production and consumption are harmonized with ecological limits.

In the realm of **agriculture**, futuristic technologies such as precision farming and vertical agriculture are revolutionizing food production while promoting sustainability. Precision agriculture utilizes data analytics, sensors, and drones to monitor crop health and optimize resource use. This targeted approach minimizes waste, reduces chemical inputs, and increases yields, contributing to food security without compromising environmental health. Vertical farming, on the other hand, enables the cultivation of crops in controlled environments, reducing land use and water consumption while promoting local food production. Both practices exemplify how technology can enhance ecological balance by aligning agricultural practices with Gaia principles.

The development of **biotechnology** also presents opportunities for supporting Gaia philosophy. Genetic engineering and synthetic biology can enhance crop resilience to climate change, pests, and diseases, reducing the need for harmful pesticides and fertilizers. By creating genetically modified organisms that require fewer resources, these technologies can help protect ecosystems and promote biodiversity. However, it is crucial to approach these innovations ethically, ensuring that they align with the principles of ecological sustainability and social equity.

Waste management technologies are another area where innovation can have a significant impact. Advances in recycling, composting, and waste-to-energy systems can reduce landfill waste and minimize environmental pollution. Technologies such as anaerobic digestion convert organic waste into biogas, providing renewable energy while also enriching soil health through nutrient-rich compost. These practices align with Gaia philosophy by closing the loop on resource use and emphasizing the importance of recycling and sustainability.

Water purification and management technologies are essential for promoting ecological health, especially in the face of increasing water scarcity. Innovations such as advanced filtration systems, desalination technologies, and rainwater harvesting techniques can help ensure access to clean water while minimizing environmental impact. By implementing these technologies, communities can enhance water resilience, protect aquatic ecosystems, and foster a more sustainable relationship with this vital resource.

The integration of **Internet of Things (IoT)** technology can further support Gaia philosophy by enhancing environmental monitoring and management. IoT devices can collect data on air and water quality, wildlife populations, and climate conditions, providing valuable insights for conservation efforts. By facilitating real-time data collection and analysis, IoT technology enables more informed decision-making and promotes proactive measures to protect ecosystems and biodiversity.

In addition to these technological advancements, **eco-design** principles are becoming increasingly important in product development. By prioritizing sustainability in design processes—considering the entire lifecycle of products from production to disposal—manufacturers can minimize waste and resource consumption. This approach aligns with Gaia philosophy by emphasizing the importance of designing products that harmonize with ecological systems and contribute to a sustainable future.

In summary, futuristic technologies have the potential to significantly advance the principles of Gaia philosophy by promoting sustainability, enhancing ecological balance, and fostering harmonious relationships between humans and the natural world. From renewable energy and precision agriculture to waste management and water purification, these innovations can contribute to a healthier planet. By embracing these technologies ethically and responsibly, individuals and communities can work toward a future that honors the interconnectedness of all life and supports the well-being of the Earth.

Conclusion: Gaia Philosophy and its Global Impact

Gaia philosophy has emerged as a significant framework for understanding the intricate relationships between living organisms and their environments, offering profound insights into the interconnectedness of life on Earth. By viewing the planet as a self-regulating system, this philosophy emphasizes the importance of ecological balance and the need for sustainable practices that honor the natural world. Its global impact is increasingly evident across various sectors, from environmental activism and policy-making to education and community initiatives.

At the core of Gaia philosophy is the recognition that all life forms are interconnected. This perspective encourages individuals and communities to adopt a holistic view of environmental issues, understanding that the health of ecosystems directly influences human well-being. As a result, Gaia-inspired initiatives promote biodiversity conservation, sustainable resource management, and ecological restoration. These efforts contribute to the preservation of natural habitats, the protection of endangered species, and the restoration of degraded ecosystems, reinforcing the idea that protecting the planet is essential for the survival of all living beings.

The principles of Gaia philosophy also resonate within the realm of **environmental policy and governance**. Policymakers increasingly recognize the need to consider ecological systems in decision-making processes. Integrated approaches that address social, economic, and environmental factors are becoming more common, as the interconnectedness emphasized by Gaia philosophy underscores the importance of sustainable development. This shift toward holistic governance can lead to more effective policies that promote long-term ecological health and resilience.

In the field of **education**, Gaia philosophy has inspired curricula that emphasize environmental literacy and stewardship. Educational programs that highlight the interdependence of life and the importance of ecological systems foster a sense of responsibility among students. By cultivating a deeper appreciation for nature, these programs empower future generations to become advocates for the planet, equipped with the knowledge and skills to address pressing environmental challenges.

Community engagement is another critical aspect of Gaia philosophy's global impact. Grassroots movements inspired by its principles encourage local action to promote sustainability and environmental justice. Initiatives such as community gardens, clean-up campaigns, and conservation projects foster a sense of connection to the land and encourage collective responsibility for the environment. By working together, communities can create meaningful change, demonstrating that local actions can have a global impact.

Furthermore, the ethical dimensions of Gaia philosophy emphasize the intrinsic value of all living beings. This perspective challenges anthropocentric views that prioritize human interests above those of the natural world. The recognition of the rights of non-human organisms fosters compassion and encourages ethical decision-making in various sectors, including business, agriculture, and resource management. As more individuals and organizations embrace this ethical framework, there is potential for transformative change in how society interacts with the environment.

The integration of **futuristic technologies** with Gaia philosophy further enhances its global impact. Innovations in renewable energy, sustainable agriculture, and waste management align with the principles of ecological balance and resource conservation. By leveraging technology to support sustainable practices, society can work toward a future that honors the interconnectedness of all life forms while addressing the pressing challenges posed by climate change and environmental degradation.

In conclusion, Gaia philosophy serves as a guiding framework for fostering a deeper understanding of the interconnectedness of life on Earth. Its global impact is evident in environmental activism, policy-making, education, community initiatives, and ethical considerations. By promoting sustainability, biodiversity conservation, and ecological stewardship, this philosophy inspires individuals and communities to engage in meaningful actions that protect the planet. As humanity navigates the complexities of the 21st century, embracing Gaia philosophy can lead to a more sustainable, equitable, and harmonious future for all living beings.

Implications for Science and Society

Gaia philosophy carries profound implications for both science and society, influencing how we understand ecological systems and our role within them. By viewing the Earth as a self-regulating organism, this perspective encourages a holistic approach to scientific inquiry and fosters a sense of collective responsibility among individuals and communities.

In the realm of **science**, Gaia philosophy has prompted a shift toward **interdisciplinary research** that recognizes the interconnectedness of biological, physical, and social systems. This holistic approach encourages scientists to collaborate across disciplines, integrating insights from ecology, climate science, sociology, and economics. For instance, research on climate change increasingly considers the feedback loops within ecosystems and the socio-economic factors that influence human behavior. This comprehensive understanding can lead to more effective strategies for addressing environmental challenges.

Moreover, Gaia philosophy has influenced the field of **ecology** by highlighting the importance of biodiversity and ecosystem health. The recognition that all species are interconnected underscores the need for conservation efforts that protect not only individual species but also the ecosystems they inhabit. This approach has led to initiatives aimed at preserving biodiversity hotspots, restoring degraded habitats, and promoting sustainable land-use practices. By emphasizing the intrinsic value of all life forms, Gaia philosophy fosters an ethical commitment to environmental stewardship within the scientific community.

In terms of **environmental science**, the principles of Gaia can enhance our understanding of ecosystem dynamics and resilience. Research that incorporates Gaia philosophy often focuses on the impacts of human activities on ecological systems, examining how changes in one part of the ecosystem can lead to cascading effects. This perspective informs conservation strategies and restoration efforts by emphasizing the need to maintain ecological balance and promote resilience in the face of disturbances such as climate change or habitat destruction.

The implications of Gaia philosophy extend beyond science to influence **societal values and behaviors**. By fostering a deeper appreciation for the interconnectedness of life, this philosophy encourages individuals to recognize their role in the larger ecological community. This awareness can lead to increased environmental activism and a commitment to sustainable practices. As people come to understand that their actions have consequences for the planet, they may be more inclined to adopt eco-friendly behaviors, support conservation efforts, and advocate for policies that prioritize ecological health.

Furthermore, Gaia philosophy encourages the integration of **sustainability** into various aspects of society, including education, economics, and governance. In education, curricula that incorporate Gaia principles can cultivate environmental literacy and inspire future generations to engage in ecological stewardship. Programs that emphasize the importance of biodiversity, ecosystem services, and sustainability foster a sense of responsibility among students, equipping them with the knowledge and skills to address pressing environmental issues.

In the economic realm, Gaia philosophy can inform the development of **sustainable business practices**. Companies that embrace this perspective recognize the importance of operating within ecological limits, prioritizing sustainable resource management, and considering the long-term impacts of their operations on the environment. This shift toward sustainable business models can lead to innovations that benefit both the economy and the planet, promoting practices such as circular economy principles, renewable energy utilization, and responsible sourcing.

At the level of governance, integrating Gaia philosophy into policymaking can lead to more effective environmental policies. This approach encourages lawmakers to consider the interconnectedness of ecological systems and the social dimensions of environmental issues. Policies that promote conservation, renewable energy, and sustainable land use can be informed by an understanding of the complex relationships within ecosystems. Engaging communities in decision-making processes can also enhance the effectiveness of policies, as local knowledge and values are integrated into environmental management strategies.

In conclusion, the implications of Gaia philosophy for science and society are profound and far-reaching. By promoting an interdisciplinary approach to research, emphasizing the importance of biodiversity and ecosystem health, and fostering a sense of interconnectedness among individuals, this philosophy has the potential to transform our understanding of ecological systems and inspire meaningful action toward sustainability. As humanity confronts the challenges of the 21st century, embracing the principles of Gaia philosophy can lead to a more resilient and harmonious relationship with the Earth, ensuring a sustainable future for all living beings.

Our Responsibility Towards Mother Earth

The concept of responsibility towards the Earth is a fundamental tenet of Gaia philosophy, which posits that our planet functions as a living, interconnected system. This perspective underscores the idea that all life forms, including humans, are part of a larger ecological community, and it emphasizes the importance of stewardship in maintaining the health and balance of our environment.

Recognizing our responsibility towards the Earth begins with understanding the **interconnectedness** of all living beings. Each organism plays a vital role in sustaining ecosystems, contributing to nutrient cycling, pollination, and other essential processes. This interdependence highlights that human actions directly affect not only our own well-being but also the health of the entire planet. As such, taking responsibility for our actions is crucial to ensuring the survival of diverse species and the ecosystems they inhabit.

One of the primary ways to fulfill this responsibility is through **sustainable practices**. Adopting eco-friendly habits, such as reducing waste, conserving water, and minimizing energy consumption, can significantly lessen our ecological footprint. For instance, using reusable products, composting organic waste, and supporting local and sustainable food sources help mitigate pollution and resource depletion. These actions reflect a commitment to living in harmony with nature and demonstrate an understanding of the impact of consumption on the environment.

Another critical aspect of our responsibility is to **advocate for conservation** and the protection of biodiversity. As human activities continue to threaten natural habitats and contribute to species extinction, it becomes essential to support efforts aimed at preserving ecosystems. This can involve participating in local conservation projects, advocating for policies that protect endangered species, and supporting organizations dedicated to environmental protection. By actively engaging in conservation, we can help restore balance to ecosystems and ensure the survival of countless species.

Education and awareness are also vital components of our responsibility towards the Earth. By informing ourselves and others about ecological issues, we can foster a greater understanding of the interconnectedness of life. This knowledge empowers individuals to make informed choices and advocate for sustainable practices within their communities.

Educational initiatives that emphasize the importance of biodiversity, ecosystem services, and sustainable living can inspire collective action and encourage a culture of environmental stewardship.

Engaging with nature is another essential way to fulfill our responsibility. Spending time outdoors fosters a deeper appreciation for the natural world and strengthens our connection to it. This relationship can inspire a sense of stewardship and motivate individuals to protect the environment. Activities such as hiking, volunteering for local clean-ups, or participating in community gardens allow individuals to engage with their surroundings while promoting ecological health.

Additionally, addressing **social and environmental justice** is an integral part of our responsibility towards the Earth. Environmental degradation often disproportionately affects marginalized communities, who may lack access to clean air, water, and green spaces. Advocating for equitable policies that ensure all individuals have a healthy environment is essential for creating a just and sustainable future. This commitment to eco-justice reinforces the understanding that the health of the planet and the well-being of its inhabitants are deeply intertwined.

Finally, fostering an **ethical mindset** towards nature encourages a profound sense of responsibility. Recognizing the intrinsic value of all living beings promotes compassion and respect in our interactions with the environment. This perspective can lead to choices that prioritize ecological health, such as supporting cruelty-free products, advocating for animal rights, and participating in conservation efforts that honor the dignity of all life forms.

In summary, embracing our responsibility towards the Earth as articulated in Gaia philosophy involves recognizing our interconnectedness with all living beings, adopting sustainable practices, advocating for conservation, educating ourselves and others, engaging with nature, addressing social and environmental justice, and fostering an ethical mindset. By fulfilling these responsibilities, we can contribute to the health and balance of our planet, ensuring a sustainable future for generations to come. Through collective action and individual commitment, we can honor the Earth as our shared home and safeguard its vitality for all life forms.

Have Questions / Comments?

This book was designed to cover as much as possible but I know I have probably missed something, or some new amazing discovery that has just come out.

If you notice something missing or have a question that I failed to answer, please get in touch and let me know. If I can, I will email you an answer and also update the book so others can also benefit from it.

Thanks For Being Awesome :)

Submit Your Questions / Comments At:

https://questions.xspurts.com

Get Another Book Free

We love writing and have produced a huge number of books.

For being one of our amazing readers, we would love to offer you another book we have created, 100% free.

To claim this limited time special offer, simply go to the site below and enter your name and email address.

You will then receive one of my great books, direct to your email account, 100% free!

https://free.xspurts.com

www.ingramcontent.com/pod-product-compliance
Lightning Source LLC
Chambersburg PA
CBHW050329230526
45471CB00005B/2414